Modern Birkhäuser Classics

Many of the original research and survey monographs in pure and applied mathematics published by Birkhäuser in recent decades have been groundbreaking and have come to be regarded as foundational to the subject. Through the MBC Series, a select number of these modern classics, entirely uncorrected, are being re-released in paperback (and as eBooks) to ensure that these treasures remain accessible to new generations of students, scholars, and researchers.

Rolf Berndt
Ralf Schmidt

Elements of the Representation Theory of the Jacobi Group

Reprint of the 1998 Edition

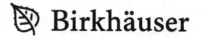 Birkhäuser

Rolf Berndt
Department Mathematik
Universität Hamburg
Bundesstr. 55
20146 Hamburg
Germany

Ralf Schmidt
Department of Mathematics
University of Oklahoma
Norman, OK 73019-3103
USA

ISBN 978-3-0348-0282-6   e-ISBN 978-3-0348-0283-3
DOI 10.1007/978-3-0348-0283-3
Springer Basel Dordrecht Heidelberg London New York

Library of Congress Control Number: 2011941499

Mathematics Subject Classification (2010): 11F55, 11F50, 11F70, 14K25, 22E50, 22E55

© Springer Basel AG 1998
Reprint of the 1st edition 1998 by Birkhäuser Verlag, Switzerland
Originally published as volume 163 in the Progress in Mathematics series

Printed on acid-free paper

Springer Basel AG is part of Springer Science+Business Media
(www.birkhauser-science.com)

Auffallender Weise hat eine so wichtige Function noch
keinen andern Namen, als den der Transcendente Θ,
nach der zufälligen Bezeichnung, mit der sie zuerst bei
J a c o b i erscheint, und die Mathematiker würden nur
eine Pflicht der Dankbarkeit erfüllen, wenn sie sich
vereinigten ihr J a c o b i s Namen beizulegen, um das
Andenken des Mannes zu ehren, zu dessen schönsten
Entdeckungen es gehört, die innere Natur und hohe
Bedeutung dieser Transcendente zuerst erkannt zu
haben.

from: L. DIRICHLET: Gedächtnisrede auf C.G.J. JACOBI

Preface

The Jacobi group is a semidirect product of a symplectic group with a Heisenberg group. Its importance prima facie stems from the fact that it sets the frame to treat theta functions and elliptic and abelian functions. Up to now, most work concerning this group has been done for the simplest case "of degree one", where the symplectic group is simply $SL(2)$ and the Heisenberg group is a three parameter nilpotent group. The Jacobi group, whose theory is intensively interwoven with that of the metaplectic group, is, together with the Heisenberg group, the most evident example for a non-reductive group. This treatise is meant to show how the general theory of automorphic forms for reductive groups extends by some slight alterations to this first more general example. The reader will see that a lot of the following may easily be extended to the higher degree case of a semidirect product of a symplectic group $Sp(n)$ with a corresponding Heisenberg group. We were tempted to do this, but as the generalizations are sometimes fairly easy on the one hand, and as the degree-one case has special features, e.g. concerning the cusp conditions, on the other hand, we restrict ourselves to this case, denoted G^J, here.

Contents

vii

Introduction

After Pyatetski-Shapiro [PS1] and Satake [Sa1] introduced, independent of one another, an early form of the Jacobi Theory in 1969 (while not naming it as such), this theory was given a definite push by the book *The Theory of Jacobi Forms* by Eichler and Zagier in 1985. Now, there are some overview articles describing the developments in the theory of the Jacobi group and its automorphic forms, for instance by Skoruppa [Sk2], Berndt [Be5] and Kohnen [Ko]. We refer to these for more historical details and many more names of authors active in this theory, which stretches now from number theory and algebraic geometry to theoretical physics. But let us only briefly indicate several – sometimes very closely related – topics touched by Jacobi theory as we see it:

- fields of meromorphic and rational functions on the universal elliptic curve resp. universal abelian variety
- structure and projective embeddings of certain algebraic varieties and homogeneous spaces
- correspondences between different kinds of modular forms
- L-functions associated to different kinds of modular forms and automorphic representations
- induced representations
- invariant differential operators
- structure of Hecke algebras
- determination of generalized Kac-Moody algebras

and as a final goal related to the here first mentioned

- mixed Shimura varieties and mixed motives.

Now, letting completely aside the arithmetical and algebraic geometrical approach to Jacobi forms developed and instrumentalized by Kramer [Kr], we will treat here a certain representation theoretic point of view for the Jacobi theory parallel to the theory of Jacquet-Langlands [JL] for GL(2) as reported by Godement [Go2], Gelbart [Ge1] and, recently, Bump [Bu]. Our text assembles and regroups material from several papers mainly on the real theory by the first-named author, and by the second-named author some definite ameliorations and additions to the non-archimedean and adelic theory contained in the thesis of Homrighausen [Ho].

More precisely, our aim is

- to give a classification of the irreducible unitary representations π of the Jacobi group G^J over local fields,

- to construct explicit models for these representations, in particular the Whittaker models,

- to discuss the relation between automorphic forms for G^J, i.e. the holomorphic Jacobi forms, their skew holomorphic counter parts (first studied by Skoruppa) and possible generalizations, and the automorphic representations of G^J, and

- to prepare the ground for a further discussion of automorphic L-functions.

To reach these aims, we pursue the following plan.

In the first chapter we present the Jacobi group G^J in some different realizations and determine its Lie algebra $\mathfrak{g}^J$. This gives some ideas about the structure of our non-reductive G^J, and indicates in particular the important subgroups of G^J one should look at. We take a closer look at the real points $G^J(\mathbb{R})$ of G^J, where we find a sort of generalized Iwasawa decomposition $G^J = N^J A^J K^J$. Here N^J is a substitute for the unipotent radical of a maximal parabolic subgroup in the reductive theory. It is characterized as the closed connected subgroup of G^J whose Lie algebra is the sum of positive root spaces.

In chapter two a method of Mackey will lead us to the fundamental principle in the representation theory of G^J, which reads in our language:

$$\boxed{\pi \simeq \tilde{\pi} \otimes \pi_{SW}^m}$$

Here π is a representation of G^J, $\tilde{\pi}$ is a genuine representation of the metaplectic group Mp, and π_{SW}^m is a certain projective standard representation of G^J, called the **Schrödinger-Weil representation**. The meaning of the above equation is that there is a 1-1 correspondence

$$\left\{ \begin{array}{l} \text{Irreducible representations} \\ \text{of } G^J \text{ with fixed non-trivial} \\ \text{central character} \end{array} \right\} \longleftrightarrow \left\{ \begin{array}{l} \text{Irreducible, genuine re-} \\ \text{presentations of Mp} \end{array} \right\} .$$

One of our objectives in the later chapters is to make the above isomorphism explicit, thereby showing that this bijection is canonical.

As the methods to be used in the archimedean, the non-archimedean and the adelic theory are sometimes quite different, we treat these cases separately starting with the real case in the chapters 3 and 4. Here the key for obtaining explicit information is a mixture of

- the induction procedure and
- the infinitesimal method realizing $\mathfrak{g}^J \otimes \mathbb{C}$ by differential operators.

On the way we discuss covariant differential operators on the homogeneous space

$$\mathbf{X} = \mathbf{H} \times \mathbb{C} = G^J(\mathbb{R})/SO(2) \times \mathbb{R}.$$

and determine the (noncommutative) ring of invariant differential operators.

Introducing Satake's determination of an automorphic factor for G^J, we discuss the definition of holomorphic and more general Jacobi forms in a first step as functions on $\mathbf{X}$ and in a second step as functions on $G^J(\mathbb{R})$ (and later on in a third and final step on the adelized group $G^J(\mathbb{A})$). Aiming at the proof of a duality theorem, tying automorphic forms of a certain type with corresponding cuspidal representations of $G^J(\mathbb{R})$, we give the definition of a cusp condition using the (conjugacy classes of) the important standard unipotent group N^J, and introduce a cuspidal subspace

$$\mathcal{H}^0 = L_0^2(\Gamma^J\backslash G^J(\mathbb{R})), \qquad \Gamma^J = \mathrm{SL}(2,\mathbb{Z}) \ltimes \mathbb{Z}^2$$

in the standard L^2-space $\mathcal{H}$, which as in the classical theory may be decomposed discretely. We observe the interesting phenomenon that in the real theory for functions ϕ with fixed transformation property with respect to the center $Z(G^J)$, i.e. with

$$\phi(\kappa g) = e^{2\pi i m\kappa}\phi(g) \qquad \text{for } \kappa \in Z(G^J),$$

the single cusp $i\infty$ of $\mathrm{SL}(2,\mathbb{Z})$ degenerates into (up to maximally) $2m$ cusps.

This chapter is concluded by a sketchy discussion of the continuous part, i. e. the orthocomplement of $\mathcal{H}^0$ in $\mathcal{H}$, where the notion of a general Jacobi Eisenstein series appears. This section in particular is open to further research looking for a better way to get at the functional equation and analytic continuation of these Eisenstein series.

The Chapters 5 and 6 treat the $\mathfrak{p}$-adic case. Here the basic ingredients are

- the induction procedure,
- the recourse to Waldspurger's results on the metaplectic group Mp, and
- information about the (local) Jacobi Hecke algebra.

In the non-archimedean case the natural objects to study are the admissible representations. The determination of all of these and the unitary representations is easy by some general results and Waldspurger's results in [Wa1]. It is more difficult to analyse which of the classes obtained are equivalent. Here we derive some results on intertwining operators, which together with an analysis of the Whittaker and Kirillov models do the task. In particular, having the adelization and application to L-functions in mind, it is of importance to discuss which representations π contain a spherical vector, meaning here an element invariant under $G^J(\mathcal{O})$, $\mathcal{O}$ the maximal order of the local field F. The determination of the spherical representations and the Hecke algebra for certain "good" cases (for the prime p in relation to the number m ruling the central character) contains the main part of Chapter 6. For an extension of these results to more general "worse" cases, we refer to the forthcoming thesis [Sch2] of R. Schmidt, University of Hamburg. Moreover, we remark that the existence of spherical vectors and of vectors of dominant weight (in the real case) allows for the definition and computation of local factors via the computation of a certain zeta integral, which is specific for our theory.

The final chapter 7 shows how the local considerations can be put together to give a global theory for the adelized group $G^J(\mathbb{A})$. Here again the Schrödinger-Weil representation π_{SW} plays a central role. In the global context it is best to realize it as a space of theta functions ϑ_f corresponding to Schwartz functions $f \in \mathcal{S}(\mathbb{A})$. The notion of automorphic representation is introduced as in the general theory, and we establish a representation theoretic analogue of a sort of "Shimura isomorphism", leading via π_{SW}^m to a one-to-one correspondence between genuine automorphic representations of the metaplectic $\mathrm{Mp}(\mathbb{A})$ and automorphic representations of $G^J(\mathbb{A})$ with fixed non-trivial central character. Classical holomorphic Jacobi forms $f \in J_{k,m}$, already characterized as functions on $G^J(\mathbb{R})$ in chapter 4, are now moreover lifted to functions on $G^J(\mathbb{A})$, and conditions characterizing these are given. The discussion of Hecke operators from chapter 6 is now enriched by a representation theoretic version of certain involutions W_p at the bad places $p|m$, acting on $J_{k,m}$ from the theory of Eichler and Zagier. The chapter culminates in a theorem stating that a Jacobi form $J_{k,m}^{\mathrm{cusp}}$, eigenform for all Hecke operators and these involutions, generates an irreducible automorphic representation π_f of $G^J(\mathbb{A})$, whose infinite component is a discrete series representation $\pi_{m,k}^+$, while for $p \nmid 2m\infty$, its p-component is a spherical principal series representation $\pi_{\chi,m}$ (characterized by the Hecke eigenvalue $c(p) = p^{k-3/2}(\chi(p) + \chi(p)^{-1})$ of f).

We have tried to write this report so it may be read independently of other texts, but we understand that some knowledge of the above mentioned sources for the GL(2)-theory will be helpful. We close this introduction by thanking several people who participated sometimes even without their knowledge. The second named author learnt a lot in courses given by S. Kudla, the first one still draws on conversations with F. Shahidi a long time ago, and with M. Eichler and D. Zagier, still a longer time ago. We both used hints and comments given by J. Michaliček, P. Slodowy and most of all by J. Dulinski. About half of the text was typeset by Mrs. D. Glasenapp, and it is our pleasure to thank her too, as well as our local TEX-adviser E. Begemann. Last not least, we appreciate the help of Mrs. C. Baer and Th. Hintermann from the Birkhäuser Verlag.

Hamburg, November 6, 1997   R. Berndt, R. Schmidt

Hanc nostram de transformatione theoriam
et, quae alia inde in analysin functionum
ellipticarum redundant, iam fusius exponemus.

from: C.G.J. JACOBI: Fundamenta Nova Theoriae Functionum Ellipticarum

1

The Jacobi Group

The Jacobi group is a semidirect product of a (semisimple) symplectic group with a (nilpotent) Heisenberg group. It comes along in several presentations which may be more or less appropriate for the different parts of the theory. So we will discuss here several realizations and change from one to the other from time to time. To keep track it is helpful to think of the Jacobi group as a certain subgroup of a bigger symplectic group.

1.1 Definition of G^J

Let R be a commutative ring with 1. Consider the symplectic group $\mathrm{Sp}(2, R)$, which is by definition the group of matrices

$$\begin{pmatrix} A & B \\ C & D \end{pmatrix} \in \mathrm{GL}(4, R),$$

where A, B, C, D are (2×2)-matrices which fulfill

$$A^t D - C^t B = E, \qquad A^t C = C^t A, \qquad B^t D = D^t B. \tag{1.1}$$

We define the **Jacobi group** $G^J(R)$ over R, our object of study, as the subgroup of $\mathrm{Sp}(2, R)$ consisting of matrices of the form

$$\begin{pmatrix} & * & \\ 0 & 0 & 0 & 1 \end{pmatrix}.$$

An easy calculation using (1.1) yields

$$\begin{pmatrix} a & 0 & b & \mu' \\ \lambda & 1 & \mu & \kappa \\ c & 0 & d & -\lambda' \\ 0 & 0 & 0 & 1 \end{pmatrix}$$

with

$$ad - bc = 1 \qquad \text{and} \qquad (\lambda, \mu) = (\lambda', \mu') \begin{pmatrix} a & b \\ c & d \end{pmatrix}$$

as the most general element of $G^J(R)$. We now identify

$$\begin{pmatrix} a & b \\ c & d \end{pmatrix} \in \mathrm{SL}(2, R) \qquad \text{with} \qquad \begin{pmatrix} a & 0 & b & 0 \\ 0 & 1 & 0 & 0 \\ c & 0 & d & 0 \\ 0 & 0 & 0 & 1 \end{pmatrix} \in G^J(R)$$

and

$$(\lambda, \mu, \kappa) \in H(R) \qquad \text{with} \qquad \begin{pmatrix} 1 & 0 & 0 & \mu \\ \lambda & 1 & \mu & \kappa \\ 0 & 0 & 1 & -\lambda \\ 0 & 0 & 0 & 1 \end{pmatrix} \in G^J(R).$$

Here $H(R)$ denotes the **Heisenberg group**, which is R^3 as a set, and with multiplication

$$(\lambda, \mu, \kappa)(\lambda', \mu', \kappa') = (\lambda + \lambda', \mu + \mu', \kappa + \kappa' + \lambda\mu' - \mu\lambda').$$

If we put $X = (\lambda, \mu)$, $X' = (\lambda', \mu')$, this can also be written as

$$(X, \kappa)(X', \kappa') = \left(X + X', \kappa + \kappa' + \begin{vmatrix} X \\ X' \end{vmatrix} \right),$$

where $| \ |$ denotes the determinant. The above identifications obviously yield injections

$$\mathrm{SL}(2, R) \hookrightarrow G^J(R) \qquad \text{and} \qquad H(R) \hookrightarrow G^J(R),$$

and it is also obvious that every element $g \in G^J(R)$ can uniquely be written as

$$g = Mh \qquad \text{or as} \qquad g = h'M'$$

with $M, M' \in \mathrm{SL}(2, R)$ and $h, h' \in H(R)$. Projection onto the $\mathrm{SL}(2)$-part is immediately recognized as a group homomorphism, yielding an exact sequence

$$1 \longrightarrow H(R) \longrightarrow G^J(R) \longrightarrow \mathrm{SL}(2, R) \longrightarrow 1. \tag{1.2}$$

This sequence splits by means of the above injection $\mathrm{SL}(2,R) \hookrightarrow G^J(R)$, so that the Jacobi group becomes the semidirect product of $\mathrm{SL}(2,R)$ and the Heisenberg group:

$$G^J(R) = \mathrm{SL}(2,R) \ltimes H(R). \tag{1.3}$$

For reasons of brevity we define $G := \mathrm{SL}(2)$. A small calculation makes the action of $G(R)$ on $H(R)$ explicit; it is given by

$$M(X,\kappa)M^{-1} = (XM^{-1},\kappa) \qquad \text{for } M \in \mathrm{SL}(2,R),\ (X,\kappa) \in H(R),$$

where XM^{-1} means matrix multiplication row times matrix. So, for example, the product of $g = M(X,\kappa)$ and $g' = M'(X',\kappa')$ is given by

$$gg' = MM'\left(XM' + X', \kappa + \kappa' + \left|\begin{matrix} XM' \\ X' \end{matrix}\right|\right), \tag{1.4}$$

and the product of $g = (X,\kappa)M$ and $g' = (X',\kappa')M'$ is given by

$$gg' = \left(X + X'M^{-1}, \kappa + \kappa' + \left|\begin{matrix} X \\ X'M^{-1} \end{matrix}\right|\right) MM'. \tag{1.5}$$

In the classic book [EZ] of Eichler and Zagier on Jacobi forms the element Mh of the (real) Jacobi group is written as a pair

$$Mh = [M,h] \qquad (M \in \mathrm{SL}(2),\ h \in H)$$

or as

$$Mh = [M,X,\kappa] \qquad (M \in \mathrm{SL}(2),\ X = (\lambda,\mu) \in R^2,\ \kappa \in R).$$

Another notation as a pair is also often used, namely

$$hM = (M,h) = (M,Y,\kappa) \qquad (M \in \mathrm{SL}(2),\ h = (Y,\kappa) \in H).$$

In case of G^J over the reals this is more than just two ways of notation, namely it results in covering the manifold $G^J(\mathbb{R})$ by two different charts. In order not to be disturbed by these different coordinatizations and notations, it may be helpful to keep in mind our first definition of G^J as a subgroup of $\mathrm{Sp}(2)$. But almost all the time the Jacobi group will be used in its realization (1.3) as a semidirect product.

1.2 G^J as an algebraic group

The Jacobi group is defined by polynomial conditions as a group of matrices, and as such it can be considered as an affine algebraic group. In order to have available all the theorems and notions for general algebraic groups, we look at $G^J(k)$, where k is an algebraically closed field of characteristic zero. But it should be noted that G^J and all of its relevant subgroups soon to be defined are already defined over $\mathbb{Q}$, so that there is no difficulty in considering the K-rational points of G^J for every field K between $\mathbb{Q}$ and k. Closer looks at the real and $\mathfrak{p}$-adic Jacobi group will be taken in subsequent chapters.

From (1.3) one can see that $G^J(k)$ is a six-dimensional closed connected subgroup of $\mathrm{GL}(4, k)$. Its Heisenberg part $H(k)$ is a unipotent group; this is seen, for instance, by realizing $H(k)$ as a group of upper triangular unipotent (3×3)-matrices via

$$\begin{pmatrix} 1 & \lambda & \kappa \\ 0 & 1 & \mu \\ 0 & 0 & 1 \end{pmatrix} \longmapsto (\lambda, \mu, 2\kappa - \lambda\mu).$$

In particular $H(k)$ contains no semisimple elements. If $G^J(k)$ would contain a two-dimensional torus, then in view of the exact sequence (1.2), the Heisenberg group would contain a nontrivial torus, which is not the case. Hence the maximal tori in $G^J(k)$ are one-dimensional. One of them is the usual $\mathrm{SL}(2)$-torus

$$A = \left\{ \begin{pmatrix} a & 0 \\ 0 & a^{-1} \end{pmatrix} : a \in k^* \right\},$$

(regarded as a subgroup of $G^J(k)$) and the others are got from A by conjugation in $G^J(k)$. Let B be the standard Borel subgroup of $\mathrm{SL}(2, k)$. Then it is immediate that

$$BH = \left\{ \begin{pmatrix} a & x \\ 0 & a^{-1} \end{pmatrix} (\lambda, \mu, \kappa) : a \in k^*, \ x, \lambda, \mu, \kappa \in k \right\}$$

is a maximal closed, connected, solvable subgroup, i.e. a Borel subgroup of $G^J(k)$. All other Borels are conjugate to this one. The unipotent radical of BH is quickly identified as

$$(BH)_u = NH = \left\{ \begin{pmatrix} 1 & x \\ 0 & 1 \end{pmatrix} (\lambda, \mu, \kappa) : x, \lambda, \mu, \kappa \in k \right\},$$

where N is the unipotent radical of B in $\mathrm{SL}(2, k)$. As usual $BH = A \ltimes (BH)_u$. We have

$$G^J(k)/BH \simeq \mathrm{SL}(2, k)/B \simeq \mathbb{P}^1(k),$$

so that $G^J(k)$ is of semisimple rank 1. In particular the Weyl group consists of two elements, the non-trivial one represented by

$$w = \begin{pmatrix} 0 & 1 \\ -1 & 0 \end{pmatrix} \in G^J(k).$$

The only Borel subgroups containing the maximal torus A are BH and its conjugate

$$wBHw^{-1} = \left\{ \begin{pmatrix} a & 0 \\ x & a^{-1} \end{pmatrix} (\lambda, \mu, \kappa) : a \in k^*, \ x, \lambda, \mu, \kappa \in k \right\}.$$

The Bruhat decomposition reads

$$G^J(k) = BH \cup BHwBH$$

(disjoint). Because of the semisimplicity of SL(2) the radical of $G^J(k)$, meaning the unique maximal closed solvable normal subgroup, is given by the Heisenberg group H. Its subgroup consisting of unipotent elements, which by definition is the unipotent radical of $G^J(k)$, is H itself, as we saw above:

$$R(G^J) = R_u(G^J) = H.$$

In particular, G^J is far from being reductive. Accordingly, its center is not a torus, but unipotent. To be more precise, a quick calculation shows that it equals the center of the Heisenberg group

$$Z = \{(0,0,\kappa) : \kappa \in k\} \simeq \mathbb{G}_a(k).$$

It will often simply be written $\kappa \in G^J(k)$ for $\kappa \in k$, meaning that κ is identified with $(0,0,\kappa) \in H \subset G^J(k)$.

The above considerations show that the standard algebraic structure of G^J is strongly dominated by the SL(2) part. The nilpotent Heisenberg part appears merely as an appendix to all the usual subgroups in SL(2). Consequently the above standard notions for algebraic groups are not best suited for working with G^J. A better insight into which subgroups should instead be considered comes from determining the Lie algebra and root structure of G^J, which will be done in the following section. But before doing this we make a remark on the derived group of G^J.

1.2.1 Proposition. *The Jacobi group is its own commutator group over any field K of characteristic not equal to 2, i.e.*

$$(G^J(K), G^J(K)) = G^J(K).$$

Proof: SL(2) is semisimple, so

$$SL(2, K) = (SL(2, K), SL(2, K)) \subset (G^J(K), G^J(K)). \tag{1.6}$$

Because of our hypothesis on the characteristic, the Heisenberg group is exactly 2-step nilpotent, more precisely

$$(H(K), H(K)) = Z(H(K)) = Z(G^J(K)),$$

where Z denotes the center. Hence

$$Z(G^J(K)) \subset (G^J(K), G^J(K)). \tag{1.7}$$

The commutator of $M \in \mathrm{SL}(2, K)$ and $(X, 0) \in H(K)$ (with $X \in K^2$) is

$$
\begin{aligned}
M(X,0)M^{-1}(-X,0) &= (XM^{-1},0)(-X,0) \\
&= \left(X(M^{-1} - 1), \left| \begin{matrix} XM^{-1} \\ -X \end{matrix} \right| \right).
\end{aligned}
$$

This together with (1.7) shows that $(G^J(K), G^J(K))$ contains all elements of the form

$$
(X(M - 1), \kappa), \qquad M \in \mathrm{SL}(2, K), \; X \in K^2, \; \kappa \in K,
$$

and this is the whole Heisenberg group. The assertion follows in view of (1.6). □

1.2.2 Corollary. *The Jacobi group has no nontrivial characters .*

1.2.3 Corollary. *The real, complex, $\mathfrak{p}$-adic and adelic Jacobi groups are all unimodular .*

Proof: The modular character is a character. □

In connection with this last corollary, we mention the following measure theoretic fact.

1.2.4 Proposition. *Consider the real, complex, $\mathfrak{p}$-adic or adelic Jacobi group. If dM and dh denote Haar measures on $\mathrm{SL}(2)$ and H, respectively, then*

$$
f \longmapsto \int\limits_{\mathrm{SL}(2)} \int\limits_{H} f(hM)\, dh\, dM = \int\limits_{\mathrm{SL}(2)} \int\limits_{H} f(Mh)\, dh\, dM
$$

(f a suitable function on G^J) defines a (biinvariant) Haar measure on G^J.

Proof: We only have to show the equality of the two integrals. Abbreviating $h^M = M^{-1}hM$, we trivially have

$$
\int\limits_{\mathrm{SL}(2)} \int\limits_{H} f(hM)\, dh\, dM = \int\limits_{\mathrm{SL}(2)} \int\limits_{H} f(Mh^M)\, dh\, dM.
$$

Our claim follows once it is shown that for fixed $M \in \mathrm{SL}(2)$ and every suitable function F on H

$$
\int\limits_{H} F(h)\, dh = \int\limits_{H} F(h^M)\, dh.
$$

But it is clear that the expression on the right also defines a Haar measure on H, and therefore the two integrals differ at most by a positive constant, which we denote by $\alpha(M)$. The map $M \mapsto \alpha(M)$ obviously is a character of $\mathrm{SL}(2)$. But this group, being semisimple, has no non-trivial characters, and we are done. □

1.3 The Lie algebra of G^J

Let k be as in the last section. The Lie algebra $\mathfrak{g}^J$ of $G^J(k)$ is very easily determined as a subalgebra of $M(4,k)$, because $G^J(k)$ was originally defined as a subgroup of $\mathrm{GL}(4,k)$, cf. Section 1.1. We just list a natural basis of $\mathfrak{g}^J$ and behind the six basis elements the closed connected subgroups of $G^J(k)$ corresponding to the one-dimensional subspaces spanned by these elements.

$$F = \begin{pmatrix} 0 & 0 & 1 & 0 \\ 0 & 0 & 0 & 0 \\ 0 & 0 & 0 & 0 \\ 0 & 0 & 0 & 0 \end{pmatrix}, \qquad \left\{ \begin{pmatrix} 1 & x \\ 0 & 1 \end{pmatrix} : x \in k \right\} \simeq \mathbb{G}_a(k).$$

$$G = \begin{pmatrix} 0 & 0 & 0 & 0 \\ 0 & 0 & 0 & 0 \\ 1 & 0 & 0 & 0 \\ 0 & 0 & 0 & 0 \end{pmatrix}, \qquad \left\{ \begin{pmatrix} 1 & 0 \\ x & 1 \end{pmatrix} : x \in k \right\} \simeq \mathbb{G}_a(k).$$

$$H = \begin{pmatrix} 1 & 0 & 0 & 0 \\ 0 & 0 & 0 & 0 \\ 0 & 0 & -1 & 0 \\ 0 & 0 & 0 & 0 \end{pmatrix}, \qquad \left\{ \begin{pmatrix} a & 0 \\ 0 & a^{-1} \end{pmatrix} : a \in k^* \right\} \simeq \mathbb{G}_m(k).$$

$$P = \begin{pmatrix} 0 & 0 & 0 & 0 \\ 1 & 0 & 0 & 0 \\ 0 & 0 & 0 & -1 \\ 0 & 0 & 0 & 0 \end{pmatrix}, \qquad \{(\lambda,0,0) : \lambda \in k\} \simeq \mathbb{G}_a(k).$$

$$Q = \begin{pmatrix} 0 & 0 & 0 & 1 \\ 0 & 0 & 1 & 0 \\ 0 & 0 & 0 & 0 \\ 0 & 0 & 0 & 0 \end{pmatrix}, \qquad \{(0,\mu,0) : \mu \in k\} \simeq \mathbb{G}_a(k).$$

$$R = \begin{pmatrix} 0 & 0 & 0 & 0 \\ 0 & 0 & 0 & 1 \\ 0 & 0 & 0 & 0 \\ 0 & 0 & 0 & 0 \end{pmatrix}, \qquad \{(0,0,\kappa) : \kappa \in k\} \simeq \mathbb{G}_a(k).$$

If K is a field between $\mathbb{Q}$ and k, then $\mathfrak{g}^J$ certainly has a K-structure, as is apparent by viewing the above matrices as elements of $M(4,K)$. The real Lie algebra could equally well have been determined analytically using the exponential function.

Because $G^J(k)$ is a semidirect product of $\mathrm{SL}(2,k)$ and $H(k)$, the Lie algebra $\mathfrak{g}^J$ is a semidirect product of $\mathfrak{sl}(2)$ and $\mathfrak{h}$, the Lie algebra of the Heisenberg group. In particular, $\mathfrak{sl}(2)$ appears as a subalgebra and $\mathfrak{h}$ as an ideal of $\mathfrak{g}^J$. The exact commutation relations fulfilled by the above basis elements are the following:

$$[F,G] = H, \qquad [H,F] = 2F, \qquad [H,G] = -2G, \qquad\qquad (1.8)$$

$$[P,Q] = 2R, \qquad [R,P] = 0, \qquad [R,Q] = 0, \qquad\qquad (1.9)$$

$$[F,P] = -Q, \qquad [F,Q] = 0, \qquad [G,P] = 0, \qquad [G,Q] = -P, \quad (1.10)$$

$$[H,P] = -P, \qquad [H,Q] = Q, \qquad [F,R] = [G,R] = [H,R] = 0. \quad (1.11)$$

The Heisenberg Lie algebra makes the Killing form on $\mathfrak{g}^J$ highly degenerated. Here is its matrix in the above basis:

	F	G	H	P	Q	R
F	0	5	0	0	0	0
G	5	0	0	0	0	0
H	0	0	10	0	0	0
P	0	0	0	0	0	0
Q	0	0	0	0	0	0
R	0	0	0	0	0	0

For comparison the matrix of the Killing form of $\mathfrak{sl}(2)$ is also given:

	F	G	H
F	0	4	0
G	4	0	0
H	0	0	8

We now come to the root space decomposition of $\mathfrak{g}^J$. The maximal torus

$$A = \begin{pmatrix} * & 0 \\ 0 & * \end{pmatrix} \subset G^J(k)$$

operates on $\mathfrak{g}^J$ via the adjoint representation. This gives the decomposition

$$\mathfrak{g}^J = \bigoplus_{n \in \mathbb{Z}} \mathfrak{g}_n^J$$

where

$$\mathfrak{g}_n^J = \left\{ X \in \mathfrak{g}^J : \forall a \in k^* \; \mathrm{Ad}\left(\begin{pmatrix} a & 0 \\ 0 & a^{-1} \end{pmatrix}\right) X = a^n X \right\}.$$

A quick calculation yields the roots $1, -1, 2, -2$. More precisely,

$$\mathfrak{g}^J = \mathfrak{g}_0^J \oplus \mathfrak{g}_1^J \oplus \mathfrak{g}_{-1}^J \oplus \mathfrak{g}_2^J \oplus \mathfrak{g}_{-2}^J,$$

with

$$\mathfrak{g}_0^J = kH \oplus kR,$$

$$\mathfrak{g}_1^J = kQ, \qquad \mathfrak{g}_{-1}^J = kP,$$

$$\mathfrak{g}_2^J = kF, \qquad \mathfrak{g}_{-2}^J = kG.$$

The roots 1 and 2 shall be called the positive ones. The picture of our one-dimensional root system is as follows:

There are two subgroups of $G^J(k)$ which will play important roles in the sequel. They are obtained by taking prominent subgroups of $SL(2, k)$ and adequately adjoining subgroups of the Heisenberg group. The first one is

$$AZ = \left\{ \begin{pmatrix} a & 0 \\ 0 & a^{-1} \end{pmatrix} (0, 0, \kappa) : a \in k^*, \kappa \in k \right\} \simeq \mathbb{G}_m(k) \times \mathbb{G}_a(k).$$

This is the subgroup of G^J belonging to $\mathfrak{g}_0^J$, and it will be for G^J what a maximal torus is for a reductive group. The second one is

$$N^J = \left\{ \begin{pmatrix} 1 & x \\ 0 & 1 \end{pmatrix} (0, \mu, 0) : x, \mu \in k \right\} \simeq \mathbb{G}_a(k)^2.$$

This N^J can be nicely characterized as the closed connected subgroup of $G^J(k)$ whose Lie algebra is the sum of the positive root spaces. Accordingly it will turn out to be something like the unipotent radical of a parabolic subgroup in a reductive group. For example, cusp forms on G^J will later on be characterized by integrating over conjugates of N^J (see section 4.2 and Definition 7.4.4).

1.4 G^J over the reals

$G^J(\mathbb{R})$, though not reductive, has several special features which allow for the application of some fairly general principles. Before going into this, we will demonstrate two slightly different ways to describe the elements of $G^J(\mathbb{R})$, where in the sequel EZ recalls the book [EZ] by Eichler and Zagier, and S points back to Siegel's parametrization of the Siegel half spaces.

The EZ-coordinates $(x, y, \theta, \lambda, \mu, \kappa)$ of an element $g \in G^J(\mathbb{R})$ are fixed as follows. For

$$g = M(X, \kappa) = [M, X, \kappa] \in G^J(\mathbb{R}),$$

we take the Iwasawa decomposition $SL(2) = NAK$ and write

$$M = \underbrace{\begin{pmatrix} 1 & x \\ 0 & 1 \end{pmatrix}}_{n(x)} \underbrace{\begin{pmatrix} y^{1/2} & 0 \\ 0 & y^{-1/2} \end{pmatrix}}_{t(y)} \underbrace{\begin{pmatrix} \cos\theta & \sin\theta \\ -\sin\theta & \cos\theta \end{pmatrix}}_{r(\theta)}$$

with $x \in \mathbb{R}$, $y \in \mathbb{R}_{>0}$, $\theta \in \mathbb{R}$ (θ is determined mod 2π only) and

$$X = (\lambda, \mu) \quad \text{with} \quad \lambda, \mu \in \mathbb{R}.$$

The S-coordinates $(x, y, \theta, p, q, \kappa)$ of $g \in G^J(\mathbb{R})$ are given by

$$g = (Y, \kappa)M = (M, Y, \kappa) \in G^J(\mathbb{R}),$$

where M is as above and

$$Y = (p, q) = XM^{-1} \in \mathbb{R}^2.$$

Using EZ-coordinates just means to covering $G^J(\mathbb{R})$ with the charts

$$(x, y, \theta, \lambda, \mu, \kappa) \longmapsto \begin{pmatrix} 1 & x \\ 0 & 1 \end{pmatrix} \begin{pmatrix} y^{1/2} & 0 \\ 0 & y^{-1/2} \end{pmatrix} r(\theta)(\lambda, \mu, \kappa),$$

and using S-coordinates amounts to cover $G^J(\mathbb{R})$ by

$$(x, y, \theta, p, q, \kappa) \longmapsto (p, q, \kappa) \begin{pmatrix} 1 & x \\ 0 & 1 \end{pmatrix} \begin{pmatrix} y^{1/2} & 0 \\ 0 & y^{-1/2} \end{pmatrix} r(\theta)$$

(the variables in appropriate parts of Euclidean space).

Sometimes it is convenient to deal with

$$G'^J(\mathbb{R}) := G^J(\mathbb{R})/\mathbb{Z},$$

and use here the central coordinate

$$\zeta = e(\kappa) \in S^1, \qquad e(u) = \exp(2\pi i u) \quad \text{for} \quad u \in \mathbb{R}.$$

Then we have parallel to the Iwasawa decomposition $SL(2) = NAK$ the decomposition

$$G'^J = N^J A^J K^J,$$

where

$$A^J = \left\{ t(y, p) := (p, 0, 0) \begin{pmatrix} y^{1/2} & 0 \\ 0 & y^{-1/2} \end{pmatrix} : p \in \mathbb{R}, \ y > 0 \right\} \simeq \mathbb{R} \rtimes \mathbb{R}_{>0},$$

and where K^J is the compact group

$$K^J = \left\{ r(\theta)\zeta : \ \theta \in \mathbb{R}, \ \zeta \in S^1 \right\}.$$

Sometimes we will extend these notations slightly to cover other cases. For instance, we will write also

$$\begin{aligned} G^J &= N^J A^J K Z, \\ g &= n(x, q)t(y, p)r(\theta)\kappa \end{aligned}$$

where here

$$n(x, q) = (0, q, 0) \begin{pmatrix} 1 & x \\ 0 & 1 \end{pmatrix}, \qquad t(y, p) = (p, 0, 0) \begin{pmatrix} y^{1/2} & 0 \\ 0 & y^{-1/2} \end{pmatrix}.$$

The EZ- and the S-coordinates are adapted to describe an action of $G^J(\mathbb{R})$ and $G'^J(\mathbb{R})$ on $\mathbf{H} \times \mathbb{C}$ in the following way: We denote

$$\tau = x + iy \in \mathbf{H} \quad \text{and} \quad z = \xi + i\eta = p\tau + q \in \mathbb{C}.$$

Then $G^J(\mathbb{R})$ and $G'^J(\mathbb{R})$ act on $\mathbf{H} \times \mathbb{C}$ by

$$g(\tau, z) := \left(M(\tau), \frac{z + \lambda\tau + \mu}{c\tau + d} \right),$$

where $g = M(X, \kappa) = [M, X, \kappa]$ is meant in the EZ-coordinates with

$$M(\tau) = \frac{a\tau + b}{c\tau + d} \quad \text{for} \quad M = \begin{pmatrix} a & b \\ c & d \end{pmatrix}.$$

This operation looks more natural in the S-coordinates. We have

$$G^J(\mathbb{R})/(\mathrm{SO}(2) \times Z(\mathbb{R})) \quad \xrightarrow{\sim} \quad \mathbf{H} \times \mathbb{C},$$
$$g = (p, q, \kappa)M \quad \longmapsto \quad g(i, 0) = (\tau, p\tau + q) \qquad (\tau = M(i)).$$

(The matrix M determines an elliptic curve, namely $\mathbb{C}/(\mathbb{Z}\tau + \mathbb{Z})$, and $Y = (p, q)$ determines the point $z = p\tau + q$ on this curve.) $G^J(\mathbb{R})$ may also be made to act on functions living on $\mathbf{H} \times \mathbb{C}$ by the aid of automorphic factors. This will be discussed in 3.4 as a side effect of the study of induced representations. For this induction procedure and the definition of theta- and zeta-transforms, we will moreover use the groups

$$\hat{N}^J := N^J Z,$$

being maximal abelian in G^J, and

$$B^J := N^J A Z.$$

As already mentioned in Corollary 1.2.3, G^J is unimodular. More precisely, from Proposition 1.2.4 and the well-known form of the invariant measures on $\mathrm{SL}(2)$ and H, we see the following.

1.4.1 Remark. Both in the EZ- and the S-coordinates a biinvariant measure on $G^J(\mathbb{R})$ is given by

$$dg = y^{-2} \, dx \, dy \, d\theta \, dp \, dq \, d\kappa.$$

B^J is not unimodular. A right-invariant measure on B^J is given by

$$d_r b = dx \, dq \, \frac{dy}{y} \, d\kappa,$$

and we have

$$d_r(b_0 b) = y_0^{3/2} d_r b.$$

1.4.2 Remark. The modular function of $B^J(\mathbb{R})$ is in S-coordinates

$$\Delta_{B^J}(b) = y^{3/2}.$$

The complexified Lie algebra $\mathfrak{g}_{\mathbb{C}}^J$ and differential operators

The EZ- and S-coordinates will be used to realize the real Lie algebra $\mathfrak{g}^J$ and its complexification $\mathfrak{g}_{\mathbb{C}}^J$ by left-invariant differential operators. Here, we have

$$\mathfrak{g}_{\mathbb{C}}^J = \mathfrak{g}^J \otimes_{\mathbb{R}} \mathbb{C} = \langle Z, X_\pm, Y_\pm, Z_0 \rangle_{\mathbb{C}}$$

with

$$Z = -i(F - G), \qquad X_\pm = \frac{1}{2}(H \pm i(F + G))$$

$$Z_0 = -iR, \qquad Y_\pm = \frac{1}{2}(P \pm iQ)$$

and thus from (1.8)–(1.11) the commutation relations $[Z_0, \mathfrak{g}_{\mathbb{C}}^J] = 0$ and

$$\begin{array}{lll} [Z, X_\pm] = \pm 2X_\pm, & [X_\pm, Y_\mp] = -Y_\pm, & [X_+, X_-] = Z, \\ [Z, Y_\pm] = \pm Y_\pm, & [X_\pm, Y_\pm] = 0, & [Y_+, Y_-] = Z_0. \end{array}$$

We put

$$\mathfrak{k}^J = \langle Z, Z_0 \rangle \qquad \text{and} \qquad \mathfrak{p}_\pm = \langle X_\pm, Y_\pm \rangle.$$

For $X \in \mathfrak{g}^J$ we define the left $G^J(\mathbb{R})$-invariant differential operator $\mathcal{L}_X$ by

$$\mathcal{L}_X \phi(g) := \frac{d}{dt} \phi(g \exp tX) \Big|_{t=0} \qquad \text{for } \phi \in \mathcal{C}^\infty(G^J(\mathbb{R})),$$

and we put

$$\mathcal{L}_X = \mathcal{L}_{X_1} + i\mathcal{L}_{X_2} \qquad \text{for } X = X_1 + iX_2 \in \mathfrak{g}_{\mathbb{C}}^J.$$

These operators do certainly not depend on the coordinates chosen, but look different in different charts. For completeness we give these operators in both types of coordinates: For $g = (Y, \zeta)M = (M, Y, \kappa)$ with the six S-coordinates $(x, y, \theta, p, q, \kappa)$ we have

$$\begin{array}{rcl} \mathcal{L}_{Z_0} &=& -i\partial_\kappa \\ \mathcal{L}_{Y_\pm} &=& (1/2)y^{-1/2}e^{\pm i\theta}(\partial_p - (x \mp iy)\partial_q - (p(x \mp iy) + q)\partial_\kappa) \\ \mathcal{L}_{X_\pm} &=& \pm(i/2)e^{\pm 2i\theta}(2y(\partial_x \mp i\partial_y) - \partial_\theta) \\ \mathcal{L}_Z &=& -i\partial_\theta \end{array}$$

and for $g = M(X, \zeta) = [M, X, \zeta]$ with $(x, y, \theta, \lambda, \mu, \kappa)$

$$
\begin{aligned}
\mathcal{L}_{Z_0} &= -i\partial_\kappa \\
\mathcal{L}_{Y_\pm} &= (1/2)(\partial_\lambda \pm i\partial_\mu \pm i(\lambda \pm i\mu)\partial_\kappa) \\
\mathcal{L}_{X_\pm} &= \pm(i/2)e^{\pm 2i\theta}(2y(\partial_x \mp i\partial_y) - \partial_\theta) + (1/2)(\lambda \pm i\mu)(\partial_\lambda \pm i\partial_\mu) \\
\mathcal{L}_Z &= -i\partial_\theta - i(-\mu\partial_\lambda + \lambda\partial_\mu).
\end{aligned}
$$

On the group $G'^J(\mathbb{R}) = G^J(\mathbb{R})/\mathbb{Z}$, one has to replace in these formulas ∂_κ by $2\pi i \zeta \partial_\zeta$.

G^J as a group of Harish-Chandra type

As announced at the beginning of this section and as already to be seen by the existence of a generalized Iwasawa decomposition, some aspects of the general theory of reductive groups may be carried over to our case. Thus, $G^J(\mathbb{R})$ is a **group of Harish-Chandra type** in the sense of Satake [Sa1] pp. 118-119. Satake discusses this in a far more general situation in his Example 2 on page 121. For our special case, we will give here and moreover in 3.4 some details adapted to our coordinates.

Satake calls a Zariski connected $\mathbb{R}$–group G with Lie algebra $\mathfrak{g}$ of **Harish–Chandra type,** if it fulfills the following two conditions.

(HC1) The complexification $\mathfrak{g}_\mathbb{C}$ of $\mathfrak{g}$ is a direct sum of vector spaces

$$\mathfrak{g}_\mathbb{C} = \mathfrak{p}_+ + \mathfrak{k}^J + \mathfrak{p}_-,$$

the "canonical decomposition", with

$$[\mathfrak{k}^J, \mathfrak{p}_\pm] \subset \mathfrak{p}_\pm, \qquad \bar{\mathfrak{p}}_+ = \mathfrak{p}_-.$$

Here $\mathfrak{p}_\pm$ are the Lie algebras of connected unipotent $\mathbb{C}$–subgroups $P_\pm$ contained in the complexification $G_\mathbb{C}$ of G.

(HC2) One has a holomorphic injection

$$P_+ \times K_\mathbb{C} \times P_- \longrightarrow P_+ K_\mathbb{C} P_- \subset G_\mathbb{C}.$$

Now, as we have in our case

$$\mathfrak{g}_\mathbb{C}^J = \langle Z, X_\pm, Y_\pm, Z_0 \rangle \qquad \text{with} \qquad [Z, Y_\pm] = \pm Y_\pm, \quad [Z, X_\pm] = \pm 2X_\pm,$$

these conditions are fulfilled for

$$\mathfrak{p}_\pm = \langle X_\pm, Y_\pm \rangle \quad \text{and} \quad \mathfrak{k}^J = \langle Z, Z_0 \rangle$$

and

$$P_+ = \left\{ (1 + zX_+)(w(1,i),0)) : z, w \in \mathbb{C} \right\},$$

$$P_- = \left\{ (1 + uX_-)(v(1,-i),0)) : u, v \in \mathbb{C} \right\},$$

$$K_\mathbb{C}^J = \left\{ R(0,0,\kappa) : R \in \mathrm{SO}(2,\mathbb{C}), \ \kappa \in \mathbb{C} \right\}.$$

A second notion is very close to this:

$\mathbf{X} = \mathbf{H} \times \mathbb{C} = G'^J(\mathbb{R})/K^J(\mathbb{R})$ as a reductive coset space

Helgason introduced (see [He1] or [He2]) for his studies of invariant differential operators and eigenspace representations the notion of a reductive coset space $\mathbf{X} = G/K$ for groups $K \subset G$ with Lie algebras $\mathfrak{k}$ and $\mathfrak{g}$ such that one has a decomposition

$$\mathfrak{g} = \mathfrak{k} + \mathfrak{m} \quad \text{with} \quad Ad_G(k)\mathfrak{m} \subset \mathfrak{m} \qquad \text{for all } k \in K.$$

Here $Ad_G(k)$ denotes the adjoint representation of G operating as usual on $\mathfrak{m}$. And this condition is fulfilled in our case again for

$$\mathfrak{k} = \langle F - G, R \rangle \qquad \text{and} \qquad \mathfrak{m} = \langle F + G, H, P, Q \rangle,$$

because we have from (1.8)–(1.11)

$$[\mathfrak{k}, \mathfrak{m}] \subset \mathfrak{m}.$$

2

Basic Representation Theory
of the Jacobi Group

Depending on whether we look at the archimedean, a $\mathfrak{p}$-adic or the adelic case, the methods for studying representations are sometimes very different. In this chapter we will collect some general material, mainly going back to Mackey, which will be useful in all three cases. We start by explaining the induction procedure, and apply it to describe the representations of the Heisenberg group. We treat the representations of the Jacobi group G^J with trivial central character and set the way for all further discussions of the cases with non-trivial central character by introducing a certain projective representation of G^J, the **Schrödinger-Weil representation** (others would perhaps call it the oscillator representation). This fundamental representation will later on be elaborated thoroughly in the different cases, and will allow to reduce, in a sense to be made precise later, the G^J-theory to the metaplectic theory.

2.1 Induced representations

There is a general method (studied in detail by Mackey) to construct representations of a locally compact group G by an induction process starting from representations of a subgroup B. As we will apply this method later on at several occasions, we sketch here this procedure following essentially Kirillov [Ki] pp. 183-184.

There are two natural realizations of an **induced representation**:

1.) in a space of vector valued functions ϕ on the group G that transform according to a given representation σ of B under left translations by elements of the group B,

2.) in a space of vector valued functions F on the coset space $\mathbf{X} = B\backslash G$.

The transition from one model to the other is sometimes a difficult task, as we will see later on.

The first realization

To describe the first realization, we will consider a closed subgroup B of G and a representation σ of B in a Hilbert space $V = V_\sigma$. We denote by $d_r g$ and $d_r b$ right Haar measure on G resp. B and by $\Delta_G(g)$ and $\Delta_B(b)$ the modular function with

$$d_r(g_0 g) = \Delta_G(g_0)d_r g$$

resp. correspondingly for $\Delta_B(b)$. Then we induce from σ a representation

$$\pi = \mathrm{ind}_B^G \sigma$$

of G given by right translation ζ on the space $\mathcal{H} = \mathcal{H}_\pi$ of measurable V_σ-valued functions ϕ on G with the two properties

i) $\qquad \phi(bg) = \left(\dfrac{\Delta_B(b)}{\Delta_G(b)} \right)^{1/2} \sigma(b)\phi(g) \qquad\qquad$ for all $b \in B$ and $g \in G$.

ii) $\qquad \displaystyle\int_{\mathbf{X}} \|\phi(s(x))\|_v^2 \, d\mu_s(x) < \infty.$

Here

$$s : \mathbf{X} = B\backslash G \to G$$

is a **Borel section** of the projection $p : G \to B\backslash G$ given by $g \mapsto Bg$. Then every $g \in G$ can uniquely be written in the form

$$g = b \cdot s(x), \qquad b \in B, \quad x \in \mathbf{X},$$

and G (as a set) can be identified with $B \times \mathbf{X}$. Under this identification, the Haar measure on G goes over into a measure equivalent to the product of a quasi-invariant measure on $\mathbf{X}$ and the Haar measure on B. More precisely, if a quasi-invariant measure μ_s on $\mathbf{X}$ is appropriately chosen, then the following equalities are valid.

$$d_r g = \frac{\Delta_G(b)}{\Delta_B(b)} d\mu_s(x) d_r b \qquad \text{and} \qquad \frac{d\mu_s(xg)}{d\mu_s(x)} = \frac{\Delta_B(b(x,g))}{\Delta_G(b(x,g))},$$

where $b(x, g) \in B$ is defined by the relation

$$s(x)g = b(x, g)s(xg).$$

If G is unimodular, i.e. $\Delta_G \equiv 1$, and if it is possible to select a subgroup K that is complementary to B in the sense that almost every element of G can uniquely be written in the form

$$g = b \cdot k, \quad b \in B, \quad k \in K,$$

then it is natural to identify $\mathbf{X} = B\backslash G$ with K and to chose s as the embedding of K in G. In this case, we have

$$dg = \Delta_B(b)^{-1} d_r b \, d_r k = d_l b \, d_r k.$$

If both G and B are unimodular (or more generally, if $\Delta_G(b)$ and $\Delta_B(b)$ coincide for $b \in B$), then there exists a G–invariant measure on $\mathbf{X} = B\backslash G$. If it is possible to extend Δ_B to a multiplicative function on G, then there exists a relatively invariant measure on $\mathbf{X}$ which is multiplied by the factor $\Delta_B(g)\Delta_G(g)^{-1}$ under translation by g.

It is a fundamental fact that $\pi = \mathrm{ind}_B^G \sigma$ is **unitary** if σ is. In this case $\mathcal{B} = \mathcal{B}_\pi$ is a Hilbert space with a G-invariant scalar product of the form

$$\langle \phi_1, \phi_2 \rangle = \int\limits_G \langle \phi_1(g), \phi_2(g) \rangle_V \, d\mu(g),$$

where the measure μ on G is such that

$$\int\limits_G \|\phi(g)\|_V^2 \, d\mu(g) = \int\limits_{\mathbf{X}} \|\phi(s(x))\|_V^2 \, d\mu_s(x)$$

holds for all $\phi \in X$.

The second realization

Using the section $s : \mathbf{X} \to G$, we associate to each $\phi \in X$ a function f on $\mathbf{X}$ defined by

$$f(x) := \phi(s(x)).$$

Obviously ϕ is uniquely determined by f and we have an isomorphism of $\mathcal{B}_\pi$ onto the space $\mathcal{B}^\pi = L^2(\mathbf{X}, \mu_s, V)$ of V-valued functions on $\mathbf{X}$ having summable square norm with respect to the measure μ_s. The problem now is to exhibit the representation operator corresponding to the right translation ρ on $\mathcal{H}_\pi$. It can be shown that we have

$$\pi(g)f(x) = A(g, x)f(xg) \qquad \text{for} \quad f \in \mathcal{H}^\pi$$

where the operator valued function $A(g,x)$ is defined by the equality

$$A(g,x) = \left(\frac{\Delta_B(g)}{\Delta_G(b)}\right)^{1/2} \sigma(b),$$

in which the element $b \in B$ is defined from the relation

$$s(x)g = bs(xg).$$

2.2   The Schrödinger representation

As an example, we will discuss the **Heisenberg group** and its **Schrödinger representation**. From now on, almost everything depends on the choice of some additive character of the underlying field. Thus we will now introduce the so-called **additive standard characters**, following [Tate], 2.2. For every prime p (including $p = \infty$) we can define a homomorphism of additive groups

$$\lambda : \mathbb{Q}_p \longrightarrow \mathbb{R}/\mathbb{Z}$$

as follows. If $\mathbb{Q}_p = \mathbb{R}$, then $\lambda(x) = -x \bmod 1$. If p is finite, then we map a Laurent series in p to its main part:

$$\lambda\Big(\sum_{i>>-\infty}^{\infty} a_i p^i \Big) = \sum_{i>>-\infty}^{-1} a_i p^i.$$

If F is a finite extension of $\mathbb{Q}_p$, then the additive standard character

$$\psi : F \longmapsto S^1$$

is defined by

$$\psi(x) = e^{-2\pi i \lambda(\mathrm{Tr}(x))},$$

where Tr is the trace mapping $F \to \mathbb{Q}_p$. Hence if $F = \mathbb{R}$, then

$$\psi(x) = e^{2\pi i x},$$

and if $F = \mathbb{C}$, then

$$\psi(x) = e^{4\pi i \mathrm{Re}(x)}.$$

Caution: Our character is precisely the *inverse* of the character defined in [Tate]. We have made our choice of characters analogous to that in the papers [Be1]-[Be6] in the real case.

For $m \in F$, the notation

$$\psi^m(x) = \psi(mx)$$

will be used throughout. From [Tate] 2.2 it is known that the map $m \mapsto \psi^m$ identifies F with its own character group. It is also important to know that if F is discrete and $\mathfrak{d}$ denotes the absolute different of F then $\mathfrak{d}^{-1}$ is the greatest ideal of F, on which ψ is trivial. In particular, if $F = \mathbb{Q}_p$, then ψ is trivial on $\mathbb{Z}_p$ and on no bigger ideal.

Now, let F be a number field, $\{\mathfrak{p}\}$ the set of places of F, and $F_\mathfrak{p}$ the completion of F at $\mathfrak{p}$. We can define a global additive character ψ of the adele ring $\mathbb{A}$ of F by

$$\psi(x) = \prod_\mathfrak{p} \psi_\mathfrak{p}(x_\mathfrak{p}) \qquad \text{for all } x = (x_\mathfrak{p})_\mathfrak{p} \in \mathbb{A},$$

where ψ_p are the local standard characters defined above. The adele ring is also self-dual via the identification $\mathbb{A} \ni m \mapsto \psi^m$ (cf. [Tate] Theorem 4.1.1). The global character thus defined has the special property that $\psi(x) = 1$ for all $x \in F$, i.e., it is a character of $\mathbb{A}/F$. Every other such character is then of the form ψ^m with $m \in F$ ([Tate] Theorem 4.1.4). We will always consider these characters in the global theory.

Returning now to local considerations, we let F be a local field of characteristic 0, and consider

$$\begin{aligned} G &= H &= \{h = (\lambda, \mu, \kappa) : \lambda, \mu, \kappa \in F\} \\ B &= B_H &= \{b = (0, \mu, \kappa) : \mu, \kappa \in F\}. \end{aligned}$$

For ψ the additive standard character of F as explained above and $m \in F^*$, let

$$\sigma(b) = \sigma(0, \mu, \kappa) = \psi^m(\kappa) = \psi(m\kappa).$$

Here we have the simplest situation, i.e. G and B are unimodular and we have the decomposition

$$H = B_H A_H \qquad \text{with} \quad A_H = \{a = (\lambda, 0, 0) : \lambda \in F\}$$

and

$$h = (\lambda, \mu, \kappa) = (0, \mu, \kappa')(\lambda, 0, 0) \qquad \text{with} \quad \kappa' = \kappa + \lambda\mu.$$

This already shows that the **first realization** of $\pi = \operatorname{ind}_B^G \sigma$ is given by right translation ρ on the space $\mathcal{H}_\pi$ of measurable $\mathbb{C}$-valued functions ϕ on H with

$$\phi(bh) = \psi(m\kappa)\phi(h) \qquad \text{for all} \quad b \in B_H \quad \text{and} \quad h \in H$$

and

$$\int_F |\phi(\lambda, 0, 0)|^2 \, d\lambda < \infty.$$

This realization is sometimes called the **Heisenberg representation**.

The restriction map $\phi \mapsto f$ given by

$$f(x) = \phi(x, 0, 0)$$

intertwines this model with the usual **Schrödinger representation** π_S^m on the space $\mathcal{H}^\pi = L^2(F)$. The prescription given above for the representation operator $A(g, x)$ here means to solve the equation

$$s(x)h = bs(xh)$$

for given x, i. e. $s(x) = (x, 0, 0)$, and $h = (\lambda, \mu, \kappa)$ by

$$b = (0, \mu, \kappa + 2x\mu + \lambda\mu).$$

This means we have for $f \in L^2(F)$ the well known formula

$$(\pi_S^m(\lambda, \mu, \kappa)f)(x) = \psi^m(\kappa + (2x + \lambda)\mu)f(x + \lambda). \tag{2.1}$$

One can see directly that π_S^m is a unitary representation. In case F is non-archimedean, it is customary to regard π_S^m as a representation on the space of smooth vectors of π_S^m, which is just the Schwartz space $\mathcal{S}(F)$.

The representation theory of the Heisenberg group is very simple, due to the following theorem which we give in both the real and the $\mathfrak{p}$-adic cases. Proofs can for instance be found in [LV], 1.3 (for the real case) and [MVW], 2.I.2, 2.I.8 (for the $\mathfrak{p}$-adic case). The notion of **smooth representation** appearing in Theorem 2.2.2 will be explained in Section 5.1.

2.2.1 Theorem. (Archimedean Stone–von Neumann theorem)

   i) *π_S^m is an irreducible unitary representation of $H(\mathbb{R})$ with central character ψ^m, and every such is isomorphic to π_S^m.*

   ii) *A unitary representation of $H(\mathbb{R})$ with central character ψ^m decomposes into a direct sum of Schrödinger representations π_S^m.*

2.2.2 Theorem. (Non-archimedean Stone–von Neumann theorem)
Let F be a $\mathfrak{p}$-adic field.

   i) *The representation π_S^m on $\mathcal{S}(F)$ is an irreducible, smooth representation of $H(F)$ with central character ψ^m, and every such is isomorphic to π_S^m.*

   ii) *A smooth representation of $H(\mathbb{R})$ with non-trivial central character ψ^m decomposes into a direct sum of Schrödinger representations π_S^m.*

It is indeed the Stone–von Neumann theorem which enables much of our treatment of the representation theory of the Jacobi group.

2.3 Mackey's method for semidirect products

The aim of the present and the following sections is to compute the unitary dual of the Jacobi group over a local field or over the adeles of a number field. We will make a distinction between the representations which have trivial central character and those which do not. In the first case we impose a general method of Mackey for determining the unitary dual of certain semidirect products. The second case can be treated more directly by using only the Stone–von Neumann theorem. In this section we begin with presenting Mackey's method in a degree of generality that suffices for our purposes.

Let G' be a locally compact topological group and H' a commutative closed normal subgroup, such that the exact sequence

$$1 \longrightarrow H' \longrightarrow G' \longrightarrow G'/H' \longrightarrow 1 \tag{2.2}$$

splits, i.e., G' is a semidirect product of $G := G'/H'$ with H':

$$G' = G \ltimes H'.$$

We wish to determine the unitary representations of G' in terms of those of G and H'. The method to be described goes back to Mackey [Ma1] and is repeated, for instance, in [Ma2], p. 77.

Assume the unitary dual $\widehat{H'}$ is known and has been given the topology of uniform convergence on compact subsets. G' operates on H' by conjugation, and this induces an operation of G' on $\widehat{H'}$:

$$
\begin{aligned}
G' \times \widehat{H'} &\longrightarrow \widehat{H'}, \\
(g, \sigma) &\longmapsto \sigma^g,
\end{aligned}
$$

where the representation σ^g is given by

$$\sigma^g(h) = \sigma(ghg^{-1}) \qquad \text{for all } h \in H'.$$

Of course, if $g \in H$, then σ^g is equivalent to σ. Hence H operates trivially on $\widehat{H'}$, and only the action of G has to be considered.

Mackey's theory does not work for arbitrary semidirect products. One has to impose a certain smoothness condition on the orbits of G' in $\widehat{H'}$. Namely it is demanded that for every G'-orbit Ω in $\widehat{H'}$ and for every $\sigma \in \Omega$ with stabilizer $G'_\sigma \subset G'$ the canonical bijection

$$G'_\sigma \backslash G' \longrightarrow \Omega$$

be a homeomorphism. If this condition is fulfilled then H' is called regularly embedded, and $G' = G \ltimes H'$ is called a regular semidirect product.

The result of Mackey is now as follows.

2.3.1 Theorem. *Let G' be a locally compact topological group and H' a closed commutative normal subgroup such that the sequence (2.2) splits. Assume that H' is of type I and regularly embedded. For every $\sigma \in \widehat{H'}$ let G'_σ the stabilizer of σ under the above action of G' on $\widehat{H'}$, and*

$$\check{G}'_\sigma = \{\tau \in \widehat{G'_\sigma} : \tau\big|_{H'} \text{ is a multiple of } \sigma\}.$$

Then the induced representation

$$\operatorname{Ind}_{G'_\sigma}^{G'} \tau$$

is irreducible for every $\tau \in \check{G}'_\sigma$, and $\widehat{G'}$ is a disjoint union

$$\widehat{G'} = \bigcup_{\widehat{H'}/G} \{\operatorname{Ind}_{G'_\sigma}^{G'} \tau : \tau \in \check{G}'_\sigma\}.$$

2.4　Representations of G^J with trivial central character

Let R be a local field of characteristic 0 ($\mathbb{R}$ and $\mathbb{C}$ included) or the ring of adeles of a number field, and let G^J be the Jacobi group over R. In this section we determine the irreducible unitary representations of G^J which have trivial central character. These representations are obviously in 1-1 correspondence with the irreducible unitary representations of the group

$$G' := G^J/Z \simeq G \ltimes H', \qquad \text{where } H' := R^2.$$

Now G' contains H' as an abelian normal subgroup which allows determination of its unitary dual by means of the method described in the last section.

The first step is to determine the irreducible unitary representations of H'. This is very easy in our case because R is self-dual. Hence the unitary dual $\widehat{H'}$ identifies with R^2 itself by associating with $(m_1, m_2) \in R^2$ the unitary character

$$\begin{aligned}
R^2 &\longrightarrow \mathbb{C}^*, \\
(\lambda, \mu) &\longmapsto \psi^{m_1}(\lambda)\psi^{m_2}(\mu).
\end{aligned}$$

G operates on H' by conjugation and thus also on $\widehat{H'}$:

$$\begin{aligned}
G \times \widehat{H'} &\longrightarrow \widehat{H'}, \\
(M, \sigma) &\longmapsto (X \mapsto \sigma(XM))
\end{aligned}$$

(XM means matrix multiplication). A small calculation shows that under the above identification $\widehat{H'} = R^2$ this operation goes over to the natural action

$$\begin{aligned}
G \times R^2 &\longrightarrow R^2, \\
(M, Y) &\longmapsto MY
\end{aligned}$$

(now think of $Y \in R^2$ as a column vector). This makes it obvious that $\widehat{H'}$ decomposes into two G-orbits, one of them consisting only of the trivial representation. As a representative for the non-trivial characters we choose

$$\Psi : H' \longrightarrow \mathbb{C},$$
$$(\lambda, \mu) \longmapsto \psi(\lambda)$$

(corresponding to the point $(1, 0) \in R^2$). The stabilizer of the trivial representation is certainly G itself, and the stabilizer of Ψ is

$$G'_\Psi = \left\{ \begin{pmatrix} 1 & 0 \\ c & 1 \end{pmatrix} (\lambda, \mu) : c, \lambda, \mu \in R \right\}.$$

Theorem 2.3.1 gives the following result, where we leave it as an exercise to check the hypotheses in this theorem.

2.4.1 Proposition. *The irreducible unitary representations of G' are exactly the following:*

    i) *The representations σ where $\sigma\big|_{H'}$ is trivial and $\sigma\big|_G$ is an irreducible unitary representation of G.*

    ii) *The representations $\mathrm{Ind}_{G'_\Psi}^{G'} \tau$, where τ runs through the irreducible unitary representation of G'_Ψ whose restriction to H' is a multiple of Ψ.*

It remains to describe more closely the representations appearing in ii). Suppose τ is an irreducible unitary representation of G'_Ψ whose restriction to H' is a multiple of Ψ. Then an element $(\lambda, \mu) \in H'$ operates by multiplication with $\psi(\lambda)$. Thus every subspace which is invariant under the matrices $\begin{pmatrix} 1 & * \\ 0 & 1 \end{pmatrix}$ is yet invariant under G'_Ψ. Hence the restriction of τ to the matrix group must be irreducible. This group being isomorphic to R itself we see that our representation is one-dimensional and the matrices act through a unitary character of R. Conversely, given such a unitary character ψ^r with $r \in R$ it is immediately checked that

$$\begin{pmatrix} 1 & c \\ 0 & 1 \end{pmatrix} (\lambda, \mu) \longmapsto \psi^r(c)\psi(\mu)$$

defines a homomorphism $G'_\Psi \to \mathbb{C}^*$. So the representations τ from which we start our induction constitute a one-parameter family indexed by $r \in R$. Putting everything together we have the following result.

2.4.2 Theorem. *The irreducible unitary representations of G^J with trivial central character are exactly the following.*

i) *The representations σ where $\sigma\big|_H$ is trivial and $\sigma\big|_G$ is an irreducible unitary representation of G.*

ii) *The representations $\operatorname{Ind}_{G'_\Psi}^{G^J} \tau_r$, where*

$$\tau_r : G'_\Psi \longrightarrow \mathbb{C}^*,$$

$$\begin{pmatrix} 1 & c \\ 0 & 1 \end{pmatrix}(\lambda,\mu) \longmapsto \psi(rc+\mu).$$

2.5   The Schrödinger-Weil representation

It will turn out in the following section that every irreducible unitary (respectively smooth) representation π of G^J with non-trivial central character can be written as a tensor product of two representations, where one factor is a certain standard representation independent of π. The present section is devoted to introducing this so-called **Schrödinger-Weil representation**, which is not really a representation of G^J but a projective one. The construction is standard and carried out in much greater generality in [We].

Let R be the real or complex numbers, a $\mathfrak{p}$-adic field, or the adele ring of a number field, and consider $G^J = G \ltimes H$ over R. The starting point is the Schrödinger representation

$$\pi_s^m : H \longrightarrow \mathrm{GL}(V)$$

with central character ψ^m, $m \in R^*$, which was discussed in Section 2.2. Now G operates on H by conjugation inside G^J in the following way:

$$\begin{aligned} G \times H &\longrightarrow H, \\ (M,h) &\longmapsto MhM^{-1} = (XM^{-1},\kappa), \qquad h = (X,\kappa),\ X \in R^2,\ \kappa \in R. \end{aligned}$$

In particular, M leaves the central part of h untouched. Hence the irreducible unitary representation

$$\begin{aligned} H &\longrightarrow \mathrm{GL}(V) \\ h &\longmapsto \pi_s^m(MhM^{-1}) \end{aligned}$$

has central character ψ^m, just like π_s^m. By the Stone–von Neumann theorem, this conjugated representation must be equivalent to π_s^m itself, i.e., there is a unitary operator

$$\pi_w^m(M) : V \longrightarrow V$$

such that

$$\pi_s^m(MhM^{-1}) = \pi_w^m(M)\pi_s^m(h)\pi_w^m(M)^{-1} \qquad \text{for all } h \in H. \qquad (2.3)$$

By Schur's lemma, $\pi_W^m(M)$ is determined up to nonzero scalars. We fix one $\pi_W^m(M)$ for each $M \in G$ arbitrarily. Now for $M_1, M_2 \in G$ we have

$$\pi_W^m(M_1)\pi_W^m(M_2)\pi_S^m(h)\pi_W^m(M_2)^{-1}\pi_W^m(M_1)^{-1}$$
$$= \pi_W^m(M_1M_2)\pi_S^m(h)\pi_W^m(M_1M_2)^{-1},$$

and again by Schur's lemma there must exist a scalar $\lambda(M_1, M_2)$ of absolute value 1 such that

$$\pi_W^m(M_1M_2) = \lambda(M_1, M_2)\pi_W^m(M_1)\pi_W^m(M_2). \tag{2.4}$$

From the associativity law in G it follows that

$$\lambda(M_1M_2, M_3)\lambda(M_1, M_2) = \lambda(M_1, M_2M_3)\lambda(M_2, M_3),$$

which just says that λ is a 2-cocycle for the trivial G-modul S^1. The freedom in multiplying the operators $\pi_W^m(M)$ with scalars of absolute value 1 amounts to changing λ by a coboundary. Hence the representation π_S^m we started with determines in a unique way an element

$$\lambda \in H^2(G, S^1).$$

From [We] or [Ku1] it is known that

- $H^2(G(R), S^1)$ is trivial if $R = \mathbb{C}$.

- $H^2(G(R), S^1)$ consists of exactly two elements if $R = \mathbb{R}$ or $R = F$ a $\mathfrak{p}$-adic field.

It is further known that λ represents the non-trivial element of $H^2(G(R), S^1)$ if R is real or $\mathfrak{p}$-adic. In [Ge2] a version of this cocycle can be found which has the property that

$$\lambda\Big|_{\mathcal{O}^* \times \mathcal{O}^*} = 1 \qquad \text{if } R \text{ is } \mathfrak{p}\text{-adic and not an extension of } \mathbb{Q}_2.$$

We will use in all that follows this cocycle in the real or $\mathfrak{p}$-adic case, $\lambda = 1$ in the complex case, and the product of the corresponding local cocycles in the adelic case. Coming back to the above notations we see that

$$M \longmapsto \pi_W^m(M)$$

is a projective representation of G on V with multiplier λ. It is called the **Weil representation** with character ψ^m. Note that π_W^m is an ordinary representation exactly in the complex case. Otherwise we can make π_W^m into an ordinary representation by going over to the **metaplectic group** Mp (also denoted $\widetilde{G}$, or Mp(R)), which is by definition the topological group extension of G by $\{\pm 1\}$ determined by the cocycle λ. In other words, as a set we have

$$\mathrm{Mp} = G \times \{\pm 1\},$$

the multiplication is defined by

$$(M, \varepsilon)(M', \varepsilon') = (MM', \lambda(M, M')\varepsilon\varepsilon'),$$

and there is an exact sequence of topological groups

$$1 \longrightarrow \{\pm 1\} \longrightarrow \mathrm{Mp} \longrightarrow G \longrightarrow 1.$$

Now the map

$$(M, \varepsilon) \longmapsto \pi_w^m(M)\varepsilon$$

obviously defines a representation of Mp in the ordinary sense. It is also called the **Weil representation**.

We put the Schrödinger and the Weil representation together and define

$$\pi_{sw}^m : G^J \longrightarrow \mathrm{GL}(V),$$
$$hM \longmapsto \pi_s^m(h)\pi_w^m(M) \qquad \text{for all } h \in H, \, M \in G.$$

The defining property (2.3) of π_w^m immediately shows that π_{sw}^m is a projective representation of G^J with multiplier λ, the latter extended canonically to G^J. It is called the **Schrödinger-Weil representation** of G^J with central character ψ^m. We give the same name to the corresponding ordinary representation of the two-fold cover $\widetilde{G^J}$ of G^J which is defined analogously to G^J. Note that there is a commutative diagram

$$
\begin{array}{ccc}
\widetilde{G^J} & \longrightarrow & \widetilde{G} \\
\downarrow & & \downarrow \\
G^J & \longrightarrow & G
\end{array}
$$

and that $\widetilde{G^J}$ identifies with the semidirect product of $\widetilde{G}$ with H.

Finally we give some explicit formulas for the Weil representation. There will be the appearence of the so-called **Weil constant**. This is a function

$$\gamma : R^* \longrightarrow S^1$$

which depends on the different cases and on the character ψ^m.

- If $R = \mathbb{C}$ then γ is the constant function 1.

- If $R = \mathbb{R}$ then

$$\gamma(a) = e^{\pi i \, \mathrm{sgn}(m)\mathrm{sgn}(a)/4}.$$

- If $R = F$ is a $\mathfrak{p}$-adic field, then

$$\gamma(a) = \lim_{n \to \infty} \int_{\omega^{-n}\mathcal{O}} \psi^m(ax^2) \, dx \, \Big/ |\dots|.$$

- If $R = \mathbb{A}$ then γ is the (well-defined) product of local Weil constants.

If the dependence on the character ψ^m is to be emphasized, we write γ_m instead of γ. Though not obvious in the non-archimedean case, the Weil constant is always an eighth root of unity (see [We] or [Sch1]).

As a further ingredient to the explicit formulas below there is the (second) **Hilbert symbol**

$$(\cdot,\cdot) :\ R^* \times R^* \longrightarrow \{\pm 1\}.$$

If R is a local field then it is defined as

$$(a,b) = 1 \quad \Longleftrightarrow \quad b \text{ is a norm from } R(\sqrt{a}).$$

In particular the Hilbert symbol is constantly 1 in the complex case. The global Hilbert symbol is defined to be the product of the local symbols. More about Hilbert symbols can be found in texts on algebraic number theory.

Now we are ready to state the explicit formulas for the Weil representation. As a model for π_S^m the Schwartz space $\mathcal{S}(R)$ is used. Then the associated Weil representation acts on the same space as follows.

$$\left(\pi_W^m \begin{pmatrix} 1 & b \\ 0 & 1 \end{pmatrix} f \right)(x) \ = \ \psi^m(bx^2)f(x). \tag{2.5}$$

$$\left(\pi_W^m \begin{pmatrix} a & 0 \\ 0 & a^{-1} \end{pmatrix} f \right)(x) \ = \ (a,-1)\gamma(a)\gamma(1)^{-1}|a|^{1/2}f(ax). \tag{2.6}$$

$$\pi_W^m \begin{pmatrix} 0 & 1 \\ -1 & 0 \end{pmatrix} f \ = \ \gamma(1)\hat{f}. \tag{2.7}$$

Here $\hat{f}$ denotes the Fourier transformation of $f \in \mathcal{S}(R)$:

$$\hat{f}(x) = |2m|^{1/2} \int_R f(y)\psi^m(2xy)\,dy.$$

The factor $|2m|^{1/2}$ normalizes the measure on R to make Fourier inversion hold:

$$\hat{\hat{f}}(x) = f(-x).$$

It is not easy to deduce the formulas (2.5)–(2.7), but it is easy to prove them. It just has to be checked that (2.3) holds with these operators, but we will not carry this out. For the real case, see [Mum] Lemma 8.2 or [LV] Section 2.5.

Assume now R to be a local field. Since the Schrödinger representation is irreducible, the Schrödinger-Weil representation is also. But if we restrict π_{SW}^m to SL$(2,R)$, i.e., we consider the Weil representation π_W^m, then from the formulas (2.5)–(2.7) we immediately find the invariant subspaces $\mathcal{S}(F)^+$ and $\mathcal{S}(F)^-$ consisting of even resp. odd Schwartz functions. Let $\pi_W^{m\pm}$ denote the subrepresentations on these spaces. They are called the **positive** (resp. **negative**) or **even** (resp. **odd**) **Weil representations**.

2.5.1 Proposition. *The positive and negative Weil representations are irreducible, and we have*

$$\pi_W^m = \pi_W^{m+} \oplus \pi_W^{m-}.$$

Between the irreducible Weil representations there are exactly the following equivalences:

$$\pi_W^{m\pm} \simeq \pi_W^{m'\pm} \qquad \Longleftrightarrow \qquad mF^{*2} = m'F^{*2}.$$

Proof: It is easy to see from (2.5) and (2.7) that the isomorphism

$$\begin{aligned} \mathcal{S}(F) &\longrightarrow \mathcal{S}(F), \\ f &\longmapsto (x \mapsto f(ax)), \end{aligned}$$

intertwines π_W^m with $\pi_W^{a^2 m}$, for any $a \in F^*$. So if $R = \mathbb{C}$, we are done. The case $R = \mathbb{R}$ will follow from our considerations in the first part of Section 3.2. For the $\mathfrak{p}$-adic case, see [MVW] 2.II.1.   □

2.6   Representations of G^J with non-trivial central character

Let again G^J be the real or $\mathfrak{p}$-adic Jacobi group. In principle Mackey's method could also be used to determine the unitary representations of G^J with non-trivial central character. Since the Heisenberg group is not commutative, one would have to check carefully the hypotheses made in [Ma1]. However, we prefer a direct method similar to the construction in [We]. The procedure is also described in Kirillov [Ki] pp. 218–219.

When dealing with the real Jacobi group, we are interested in unitary representations, while for the $\mathfrak{p}$-adic Jacobi group, we consider smooth representations. Both cases can be treated in a very similar way. The decisive point is to have the Stone–von Neumann theorem at hand. We treat the unitary case and leave the minor changes for the $\mathfrak{p}$-adic case to the reader.

So let π be a unitary representation of the real Jacobi group G^J on a Hilbert space V with central character ψ^m, $m \neq 0$. The restriction of π to the Heisenberg group decomposes into unitary representations, each of which must be equivalent to the Schrödinger representation π_S^m with central character ψ^m, by the Stone–von Neumann theorem 2.2.1. So this restriction is isotypical, and consequently we may assume that V is a Hilbert tensor product

$$V = V_1 \otimes V_2,$$

where H acts trivially on V_1 and where V_2 is a representation space for π_S^m.

From the defining property (2.3) of π_w^m, which also acts on V_2, it follows easily that

$$\pi(M^{-1})(\mathbf{1}_{V_1} \otimes \pi_w^m(M))\pi(h) = \pi(h)\pi(M^{-1})(\mathbf{1}_{V_1} \otimes \pi_w^m(M)),$$

i.e., the operator $\pi(M^{-1})(\mathbf{1}_{V_1} \otimes \pi_w^m(M))$ commutes with the action of the Heisenberg group. Hence it must be of the form

$$\pi(M^{-1})(\mathbf{1}_{V_1} \otimes \pi_w^m(M)) = \tilde{\pi}(M) \otimes \mathbf{1}_{V_2} \qquad \text{with } \tilde{\pi}(M) \in \operatorname{Aut}(V_1).$$

As a result we were able to separate the action of G in one on V_1 and one on V_2:

$$\pi(M) = \tilde{\pi}(M) \otimes \pi_w^m(M). \tag{2.8}$$

More generally, for every element $g = hM$ of the Jacobi group with $M \in G$ and $h \in H$ we have

$$\pi(hM) = \tilde{\pi}(M) \otimes \pi_{sw}^m(hM),$$

where π_{sw}^m is the Schrödinger-Weil representation introduced in the last chapter. From (2.4) it follows that for $M_1, M_2 \in G$

$$\tilde{\pi}(M_1 M_2) = \lambda(M_1, M_2)^{-1}\tilde{\pi}(M_1)\tilde{\pi}(M_2).$$

In other words, $\tilde{\pi}$ and π_{sw}^m are both projective representations of G resp. G^J with multiplier λ^{-1} resp. λ. After tensorizing the cocycles cancel and the result is an ordinary representation of G^J. Summarizing we obtain the following result.

2.6.1 Theorem. *The above construction gives a 1-1 correspondence*

$$\tilde{\pi} \longmapsto \tilde{\pi} \otimes \pi_{sw}^m$$

between the irreducible unitary projective representations of $\mathrm{SL}(2, \mathbb{R})$ *with multiplier* λ *and the irreducible unitary representations of* $G^J(\mathbb{R})$ *with non-trivial central character* ψ^m.

The corresponding non-archimedean result is as follows.

2.6.2 Theorem. *Let F be a $\mathfrak{p}$-adic field. There is a 1-1 correspondence*

$$\tilde{\pi} \longmapsto \tilde{\pi} \otimes \pi_{sw}^m$$

between the irreducible smooth projective representations of $\mathrm{SL}(2, F)$ *with multiplier* λ *and the irreducible smooth representations of* $G^J(F)$ *with non-trivial central character* ψ^m.

The only difference in the complex case is that π_{SW}^m is a representation, not a projective one. Then $\tilde\pi$ will also turn out to be a representation of G, and we get the following result:

2.6.3 Theorem. *The map*

$$\tilde\pi \longmapsto \tilde\pi \otimes \pi_{SW}^m$$

establishes a 1-1 correspondence between irreducible, unitary representation of $SL(2,\mathbb{C})$ *and irreducible, unitary representations of* $G^J(\mathbb{C})$ *with central character* ψ^m *(* $m \in \mathbb{C}^*$ *).*

We refer the reader to Knapp [Kn] II, §4, for a classification of the irreducible, unitary representations of $SL(2,\mathbb{C})$, and thus for a classification of irreducible, unitary representations of $G^J(\mathbb{C})$.

Much more will be said in the following chapters about the correspondence $\tilde\pi \mapsto \tilde\pi \otimes \pi_{SW}^m$, with specific reference to the underlying field.

3

Local Representations: The Real Case

Here we rearrange and extend material from [Be1]–[Be4] and [BeBö]. By the general theory from the last section, we have as a fundamental object the **Schrödinger-Weil representation** π_{SW}^m which is a genuine representation of the metaplectic cover $\widetilde{G}^J(\mathbb{R})$ and may be identified with a projective representation of $G^J(\mathbb{R})$. If we tensorize π_{SW}^m with another genuine representation $\tilde{\pi}$ of the metaplectic cover $\mathrm{Mp}(\mathbb{R})$ (again to be identified with a projective representation of $\mathrm{SL}_2(\mathbb{R})$) we get

$$\pi = \pi_{SW}^m \otimes \tilde{\pi},$$

a representation of $G^J(\mathbb{R})$ with central character ψ^m, i.e.

$$\pi(0,0,\kappa) = \psi^m(\kappa) = e^{2\pi i m\kappa} \qquad \text{for all } \kappa \in \mathbb{R}.$$

This way, we get all unitary representations π with $m \neq 0$ if we take all unitary representations $\tilde{\pi}$ of $\mathrm{Mp}(\mathbb{R})$. The representations of the metaplectic group were studied to a large extent by Waldspurger [Wa1-3] and Gelbart [Ge2]. Thus, at least for the unitary representations, we easily get a rather complete picture simply by applying Mackey's method. But, since in the real theory we also have the possibility to apply the infinitesimal method, we use it here as the starting point. Afterwards we will discuss several features of the induction procedure, coming up, among other things, with the canonical automorphic factor and invariant differential operators on $\mathbf{H} \times \mathbb{C}$.

3.1   Representations of $\mathfrak{g}_{\mathbb{C}}^J$

We have already dealt with the complexified Lie algebra of the Jacobi group in Section 1.4. It is given by

$$\mathfrak{g}_{\mathbb{C}}^J = \langle Z, X_\pm \rangle + \langle X_\pm, Z_0 \rangle = \mathfrak{k}^J + \mathfrak{p}^+ + \mathfrak{p}^-,$$

where

$$\mathfrak{k}^J = \langle Z, Z_0 \rangle, \qquad \mathfrak{p}^\pm = \langle X_\pm, Y_\pm \rangle.$$

From the commutation relations given in Section 1.4, we repeat the following:

$$[Z_0, \mathfrak{g}_{\mathbb{C}}^J] = 0, \qquad [Z, Y_\pm] = \pm Y_\pm, \qquad [Z, X_\pm] = \pm 2 X_\pm.$$

Because of this decomposition, for each representation $\hat{\pi}$ of $\mathfrak{g}_{\mathbb{C}}^J$ the representation space V decomposes as

$$V = \sum_{k \in \mathbb{Z}} V_k \quad \text{with} \quad \begin{array}{ll} \hat{\pi}(Z_0)V_k = \mu V_k, & \hat{\pi}(Y_\pm)V_k \subset V_{k\pm 1}, \\ \hat{\pi}(Z)V_k = \rho_k V_k, & \hat{\pi}(X_\pm)V_k \subset V_{k\pm 2}, \end{array}$$

where μ and ρ_k are complex numbers.

3.1.1 Remark. $\mu \neq 0$ will be fixed here through out. As we will only be interested in representations $\hat{\pi}$ being a derived representation of a (unitary) representations π or $\tilde{\pi}$ of G^J resp. $\tilde{G}^J$, μ will be thought of the form

$$\mu = 2\pi m, \qquad m \in \mathbb{R}^* \text{ the "\textbf{index}" of } \pi \text{ resp. } \hat{\pi},$$

and ρ_k, the **weight** of V_k, should be an integer or a half integer. As later on we will be interested in representations of $G'^J = G^J/\mathbb{Z}$, the real number m will then be fixed as a non-zero integer.

3.1.2 Definition. *Let $\hat{\pi}$ be a representation of $\mathfrak{g}_{\mathbb{C}}^J$ with space $V = \sum V_k$ as above.*

i) $\hat{\pi}$ is called of **lowest (highest) weight** *k, if there is a $V_k \neq \{0\}$ with*

$$\hat{\pi}(\mathfrak{p}^-)V_k = \{0\} \qquad resp. \qquad \hat{\pi}(\mathfrak{p}^+)V_k = \{0\}.$$

The elements in V_k will then be called lowest (highest) weight vectors.

ii) $\hat{\pi}$ is called **spherical (nearly spherical)**, *if there is a $V_k \neq \{0\}$ with*

$$\rho_k = 0 \qquad resp. \qquad \rho_k = 1/2 \text{ or } 1.$$

Elements $v \in V_k$ will correspondingly be called spherical (nearly spherical) vectors.

If there is no danger of confusion, we will use abbreviations like

$$X = \hat{\pi}(X) \qquad \text{for } X \in \mathfrak{g}_{\mathbb{C}}^J$$

or

$$X^2 = (\hat{\pi}(X))^2,$$

the latter being understood as the operator belonging to the element X^2 in the universal enveloping algebra $U(\mathfrak{g}_{\mathbb{C}}^J)$ of $\mathfrak{g}_{\mathbb{C}}^J$. In particular, we will use the elements

$$D_\pm := X_\pm \pm (2\mu)^{-1} Y_\pm^2 \in U(\mathfrak{g}_{\mathbb{C}}^J).$$

Interpreted as operators, D_+ will later be recognized as the "heat operator".

From the general theory of the last chapter we know that there is a 1-1 correspondence between irreducible, unitary, genuine representations $\tilde{\pi}$ of Mp and irreducible, unitary representations π of G^J with central character ψ^m through the relation

$$\pi = \tilde{\pi} \otimes \pi_{SW}^m.$$

By differentiating, this remains true on the infinitesimal level. More generally, we have a bijection between irreducible representations $\tilde{\pi}$ of $\mathfrak{sl}_2$ (the Lie algebra of Mp) and irreducible representations $\hat{\pi}$ of $\mathfrak{g}_{\mathbb{C}}^J$, given by

$$\hat{\pi} = \tilde{\pi} \otimes \hat{\pi}_{SW}^m. \tag{3.1}$$

3.1.3 Remark. By more thoroughly analyzing the infinitesimal situation, we would avoid the recourse to Mackey's theory, and thereby also arrive at the correspondence (3.1).

For the representation $\hat{\pi}_{SW}^m = d\pi_{SW}^m$ of $\mathfrak{g}_{\mathbb{C}}^J$ we have the following result.

3.1.4 Proposition. Let $m \in \mathbb{R}^*$. If $m > 0$, the infinitesimal Schrödinger-Weil representation $\hat{\pi}_{SW}^m$ is a lowest weight representation. It operates on the space $V = \langle v_j \rangle_{j \in \mathbb{N}_0}$ by

$$Z_0 v_j = \mu v_j \qquad\qquad Z v_j = \left(j + \frac{1}{2}\right) v_j$$

$$Y_+ v_j = v_{j+1} \qquad\qquad X_+ v_j = -\frac{1}{2\mu} v_{j+2} \tag{3.2}$$

$$Y_- v_j = -\mu j v_{j-1} \qquad\qquad X_- v_j = \frac{\mu}{2} j(j-1) v_{j-2}$$

where $\mu = 2\pi m$ and $v_{-1} = v_{-2} = 0$ understood. If $m < 0$, then π_{SW}^m is a highest weight representation with space $V = \langle v_{-j} \rangle_{j \in \mathbb{N}_0}$, the action given by

$$Z_0 v_{-j} = \mu v_{-j} \qquad\qquad Z v_{-j} = -\left(j + \frac{1}{2}\right) v_{-j}$$

$$Y_- v_{-j} = v_{-(j+1)} \qquad\qquad X_- v_{-j} = \frac{1}{2\mu} v_{-(j+2)} \qquad (3.3)$$

$$Y_+ v_{-j} = \mu j v_{-(j-1)} \qquad\qquad X_+ v_{-j} = -\frac{\mu}{2} j(j-1) v_{-(j-2)}$$

(with $v_1 = v_2 = 0$ understood).

This will be proved in the next section. In particular, the weights (eigenvalues) of Z are given by half integers. It remains to describe the representations $\tilde{\pi}$. Since we are only interested in those representations $\hat{\pi}$ where Z has integral weights, we only have to classify those $\tilde{\pi}$ where Z acts by half integers. These representations, which we call *genuine*, were thoroughly studied by Waldspurger [Wa1] (see in particular p. 22). Taking over as far as possible here his notations, we have:

A) The **principal series representations**

$$\tilde{\pi} = \hat{\pi}_{s,\nu}, \qquad\quad s \in \mathbb{C} \setminus \{\mathbb{Z} + 1/2\}, \ \nu = \pm 1/2,$$

are given by

$$Z w_l = \left(l - \frac{1}{2}\right) w_l, \qquad\quad X_{\pm} w_l = \frac{1}{2}\left(s + 1 \pm \left(l - \frac{1}{2}\right)\right) w_{l \pm 2},$$

acting on

$$W_{s,\nu} = \langle w_l \rangle, \qquad l \in 2\mathbb{Z} + \nu + 1/2.$$

B) The **discrete series representations**

$$\tilde{\pi} = \hat{\pi}_{k_0}^{\pm}, \qquad\qquad k_0 \in \mathbb{Z} + 1/2,$$

are given by

$$\begin{aligned}
Z w_{\pm l} &= \pm(k_0 + l) w_{\pm l} \\
X_{\pm} w_{\pm l} &= w_{\pm(l+2)} \\
X_{\mp} w_{\pm l} &= -\frac{l}{2}\left(k_0 + \frac{l}{2} - 1\right) w_{\pm(l-2)}
\end{aligned}$$

acting on

$$W_{k_0}^{\pm} = \langle w_{\pm l} \rangle, \qquad l \in 2\mathbb{N}_0.$$

3.1.5 Remark. $\hat{\pi}_{k_0}^+$ is a lowest weight representation of lowest weight k_0, while $\hat{\pi}_{k_0}^-$ is a highest weight representation of highest weight $-k_0$.

Tensorizing, the following types of representations of $\mathfrak{g}_\mathbb{C}^J$ already discussed in [Be3,4] and [BeBö] appear. We give the explicit formulas only for $m > 0$ and leave it to the reader to write down the other case explicitly.

3.1.6 Proposition. For any $m > 0$, the **principal series representation**

$$\hat{\pi}_{m,s,\nu} := \hat{\pi}_{sw}^m \otimes \hat{\pi}_{s,\nu}, \qquad s \in \mathbb{C} \setminus \{\mathbb{Z}+1/2\},\ \nu = \pm 1/2,$$

acts on

$$V_{m,s,\nu} := \langle v_j \otimes w_l \rangle, \qquad j \in \mathbb{N}_0,\ l \in 2\mathbb{Z} + \nu + 1/2,$$

by

$$
\begin{aligned}
Z_0(v_j \otimes w_l) &= \mu v_j \otimes w_l \\
Y_+(v_j \otimes w_l) &= v_{j+1} \otimes w_l \\
Y_-(v_j \otimes w_l) &= -\mu j v_{j-1} \otimes w_l \\
Z(v_j \otimes w_l) &= (j+l)v_j \otimes w_l \\
X_+(v_j \otimes w_l) &= -\frac{1}{2\mu}v_{j+2} \otimes w_l + \frac{1}{2}\left(s+1+\left(l-\frac{1}{2}\right)\right)v_j \otimes w_{l+2} \\
X_-(v_j \otimes w_l) &= \frac{1}{2}\mu j(j-1)v_{j-2} \otimes w_l + \frac{1}{2}\left(s+1-\left(l-\frac{1}{2}\right)\right)v_j \otimes w_{l-2}
\end{aligned}
$$

$(\mu = 2\pi m)$. There are similar formulas for $m < 0$, using the equations (3.3) from Proposition 3.1.4.

This representation has as a **cyclic vector** the element $v_0 \otimes w_{1/2+\nu}$ characterized by

$$
\begin{aligned}
Z_0(v_0 \otimes w_{1/2+\nu}) &= \mu(v_0 \otimes w_{1/2+\nu}) \\
Z(v_0 \otimes w_{1/2+\nu}) &= \left(\frac{1}{2}+\nu\right)(v_0 \otimes w_{1/2+\nu}) \\
Y_-(v_0 \otimes w_{1/2+\nu}) &= 0 \\
(D_-D_+)(v_0 \otimes w_{1/2+\nu}) &= \frac{1}{4}\left(s^2-(\nu+1)^2\right)(v_0 \otimes w_{1/2+\nu})
\end{aligned}
$$

with the "heat operators"

$$D_\pm = X_\pm \pm \frac{1}{2\mu}Y_\pm^2$$

defined above.

3.1.7 Proposition. For any $m > 0$, the **discrete series representation**

$$\hat{\pi}_{m,k}^+ = \hat{\pi}_{sw}^m \otimes \hat{\pi}_{k_0}^+, \qquad k = k_0 + 1/2 \in \mathbb{Z},$$

acts on

$$V^+_{m,k} = \langle v_j \otimes w_l \rangle, \qquad j \in \mathbb{N}_0,\ l \in 2\mathbb{N}_0,$$

by

$$
\begin{aligned}
Z_0(v_j \otimes w_l) &= \mu v_j \otimes w_l \\
Y_+(v_j \otimes w_l) &= v_{j+1} \otimes w_l \\
Y_-(v_j \otimes w_l) &= -\mu j v_{j-1} \otimes w_l \\
Z(v_j \otimes w_l) &= (j + l + k)v_j \otimes w_l \\
X_+(v_j \otimes w_l) &= -\frac{1}{2\mu}v_{j+2} \otimes w_l + v_j \otimes w_{l+2} \\
X_-(v_j \otimes w_l) &= \frac{\mu}{2}j(j-1)v_{j-2} \otimes w_l - \frac{l}{2}\left(k - \frac{3}{2} + \frac{l}{2}\right)v_j \otimes w_{l-2}
\end{aligned}
$$

and

$$\hat{\pi}^-_{m,k} = \hat{\pi}^m_{sw} \otimes \hat{\pi}^-_{k_0}, \qquad k = k_0 + 1/2 \in \mathbb{Z},$$

acts on

$$V^-_{m,k} = \langle v_j \otimes w_{-l} \rangle, \qquad j \in \mathbb{N}_0,\ l \in 2\mathbb{N}_0,$$

by

$$
\begin{aligned}
Z_0(v_j \otimes w_{-l}) &= \mu v_j \otimes w_{-l} \\
Y_+(v_j \otimes w_{-l}) &= v_{j+1} \otimes w_{-l} \\
Y_-(v_j \otimes w_{-l}) &= -\mu j v_{j-1} \otimes w_{-l} \\
Z(v_j \otimes w_{-l}) &= (j - l + 1 - k)v_j \otimes w_{-l} \\
X_+(v_j \otimes w_{-l}) &= -\frac{1}{2\mu}v_{j+2} \otimes w_{-l} + \frac{l}{2}\left(\frac{1}{2} - k - \left(\frac{l}{2} - 1\right)\right)v_j \otimes w_{-(l-2)} \\
X_-(v_j \otimes w_{-l}) &= \frac{\mu}{2}j(j-1)v_{j-2} \otimes w_{-l} + v_j \otimes w_{-(l+2)}
\end{aligned}
$$

($\mu = 2\pi m$). There are similar formulas for $m < 0$, using the equations (3.3) from Proposition 3.1.4.

There is in both cases a **cyclic vector** $v_0 \otimes w_0$ of "dominant weight" characterized by

$$\hat{\pi}(Z_0) = \mu$$

and

$$\hat{\pi}(Z)v_0 \otimes w_0 = k\,v_0 \otimes w_0, \qquad \hat{\pi}(Y_-)v_0 \otimes w_0 = \hat{\pi}(X_-)v_0 \otimes w_0 = 0$$

for $\hat{\pi}^+_{m,k}$ resp.

$$\hat{\pi}(Z)v_0 \otimes w_0 = (1 - k)v_0 \otimes w_0, \qquad \hat{\pi}(Y_-)v_0 \otimes w_0 = \hat{\pi}(D_+)v_0 \otimes w_0 = 0$$

for $\pi^-_{m,k}$.

3.1.8 Remark. There is a slight asymmetry in the naming of $\mathfrak{g}^J$-representations which comes from the fact that $\hat{\pi}_{SW}^m$ raises the weight by $1/2$ (if $m > 0$): The distinguished vector of $\hat{\pi}_{m,k}^+$ has weight k, while in $\hat{\pi}_{m,k}^-$ it has weight $1 - k$. One might therefore feel the temptation to index our $\hat{\pi}_{m,k}^+$ by another integer instead of k, for example by $1 - k$ or by $k - 1$. We have thought about this problem for hours, and come to the conclusion that the choice made offers some convincing advantages. For instance, the formula for the eigenvalue of the Casimir operator C given below in Proposition 3.1.10 is the same for $\hat{\pi}_{m,k}^+$ and $\hat{\pi}_{m,k}^-$. Another, and perhaps more important, point is that, as will be seen in Section 4.1, the Jacobi forms in $J_{k,m}$ resp. $J_{k,m}^*$ correspond to representations $\pi_{m,k}^+$ resp. $\pi_{m,k}^-$.

Taking into account the above mentioned classification of genuine metaplectic representations, we arrive at the following **classification of infinitesimal representations** of G^J (which is in fact a classification of irreducible $(\mathfrak{g}_{\mathbb{C}}^J, K)$-modules, $K = SO(2)$, though we have not mentioned this terminology).

3.1.9 Theorem. *Let $m \in \mathbb{R}^*$. The following is a complete list of the irreducible representations of $\mathfrak{g}_{\mathbb{C}}^J$ where Z_0 acts by $\mu = 2\pi m$ and Z has integral weights.*

i) The **principal series representations**

$$\hat{\pi}_{m,s,\nu} := \hat{\pi}_{SW}^m \otimes \hat{\pi}_{s,\nu}$$

for $s \in \mathbb{C} \setminus \{\mathbb{Z} + 1/2\}$, $\nu = \pm 1/2$.

ii) The **positive discrete series representations**

$$\hat{\pi}_{m,k}^+ = \hat{\pi}_{SW}^m \otimes \hat{\pi}_{k_0}^+$$

for $k = k_0 + 1/2 \in \mathbb{Z}$.

iii) The **negative discrete series representations**

$$\hat{\pi}_{m,k}^- = \hat{\pi}_{SW}^m \otimes \hat{\pi}_{k_0}^-$$

for $k = k_0 + 1/2 \in \mathbb{Z}$.

The only equivalences between these representations are

$$\hat{\pi}_{m,s,\nu} \simeq \hat{\pi}_{m,-s,\nu},$$

all other representations are inequivalent.

In the following section we will decide which of these representations are unitarizable, thereby classifying the irreducible, unitary representations of $G^J(\mathbb{R})$.

There is another somewhat different approach to the determination of the representations of $\mathfrak{g}^J$ proposed by Borho and exploited in [BeBö]: If the universal enveloping algebra $U(\mathfrak{g}_{\mathbb{C}}^J)$ is localized to $U(\mathfrak{g}_{\mathbb{C}}^J)'$ by dividing out the principal ideal generated by $Z_0 - \mu$, there is a Lie homomorphism

$$' : \mathfrak{sl}_2 \longrightarrow U(\mathfrak{g}_{\mathbb{C}}^J)' := U(\mathfrak{g}_{\mathbb{C}}^J)/(Z_0 - \mu)$$

given by

$$
\begin{aligned}
X_+ &\mapsto D_+ := X_+ + (2\mu)^{-1}Y_+^2 \\
X_- &\mapsto D_- := X_- - (2\mu)^{-1}Y_-^2 \\
Z &\mapsto \Delta_1 := Z + (2\mu)^{-1}\Delta_0, \qquad \Delta_0 := Y_+Y_- + Y_-Y_+
\end{aligned}
$$

(this can be verified by direct calculation or seen from [Bo] Lemma 3.4). As a consequence of the relation $[\mathfrak{h}, \mathfrak{sl}_2'] = 0$ in $U(\mathfrak{g}_{\mathbb{C}}^J)'$ (which is easy to check), there is an isomorphism

$$i : U(\mathfrak{h}_{\mathbb{C}})' \otimes U(\mathfrak{sl}_2') \xrightarrow{\sim} U(\mathfrak{g}_{\mathbb{C}}^J)', \qquad U(\mathfrak{h}_{\mathbb{C}})' = U(\mathfrak{h}_{\mathbb{C}})/(Z_0 - \mu).$$

which (also) gives an explanation that the representations $\hat{\pi}$ of $\mathfrak{g}_{\mathbb{C}}^J$ with $\hat{\pi}(Z_0) = \mu = 2\pi m \neq 0$ are of the type

$$\pi = \pi^m \otimes \tilde{\pi}', \quad V = V_m \otimes W',$$

where $\mathfrak{h}_{\mathbb{C}}$ acts as usual on $V_m = \langle \xi_j \rangle_{j \in \mathbb{N}_0}$ by

$$Z_0\xi_j = \mu\xi_j, \qquad Y_+\xi_j = \xi_{j+1}, \quad Y_-\xi_j = -\mu j\xi_{j-1}$$

and $\tilde{\pi}'$ is one of the representations given above in A) and B) below 3.1.4, but now thought of as representations of $\mathfrak{sl}_2'$ resp. as $U(\mathfrak{sl}_2')$-modules. In particular, this explains nicely the appearance of the "heat operators" $D_\pm$ and shows that

$$C := D_+D_- + D_-D_+ + (1/2)\Delta_1^2$$

is a **Casimir operator** for the representations $\hat{\pi}$ of $\mathfrak{g}_{\mathbb{C}}^J$ with $\hat{\pi}(Z_0) = \mu \neq 0$.

3.1.10 Proposition. *The image of the operator C lies in the center of $U(\mathfrak{g}_{\mathbb{C}}^J)'$. Consequently C acts on the irreducible representations of $\mathfrak{g}_{\mathbb{C}}^J$ given in Theorem 3.1.9 by multiplication with a scalar λ. We have*

$$
\lambda = \begin{cases} \dfrac{1}{2}(s^2 - 1) & \text{for } \hat{\pi}_{m,s,\nu}, \\[2mm] \dfrac{1}{2}\left(k - \dfrac{1}{2}\right)\left(k - \dfrac{5}{2}\right) & \text{for } \hat{\pi}_{m,k}^{\pm}. \end{cases}
$$

Proof: These are straightforward calculations. □

3.2 Models for infinitesimal representations and unitarizability

In this section we present models for the infinitesimal Schrödinger-Weil representation as well as for the principal and discrete series representations of the last section, and after that discuss the question of unitarizability of these representations.

The infinitesimal Schrödinger-Weil representation

We want to compute and characterize the derived representation of π_{SW}^m on the Lie algebra $\mathfrak{g}^J$, acting on the space of smooth vectors $\mathcal{S}(\mathbb{R}) \subset L^2(\mathbb{R})$. We have already used the result in Proposition 3.1.4. Similar formulas like the ones in Proposition 3.2.1 and 3.2.2 below also appear in Section 2.5 of [LV].

The two-fold cover $\widetilde{G^J}$ is a real Lie group, and the exact sequence

$$1 \longrightarrow \{\pm 1\} \longrightarrow \widetilde{G^J} \longrightarrow G^J \longrightarrow 1$$

yields an isomorphism of Lie algebras

$$\widetilde{\mathfrak{g}^J} \overset{\sim}{\longrightarrow} \mathfrak{g}^J.$$

We conclude from this that the infinitesimal representation

$$d\pi_{SW}^m : \mathfrak{g}^J \longrightarrow \mathfrak{gl}(\mathcal{S}(\mathbb{R}))$$

is really a homomorphism of Lie algebras, and the projectivity of π_{SW}^m is no longer visible on the infinitesimal level. For an element $X \in \mathfrak{g}^J$ the operator $d\pi_{SW}^m(X)$, often simply written as X, is given by

$$(d\pi_{SW}^m(X)f)(x) = \frac{d}{dt}\Big((\exp(tX)f)(x)\Big)\Big|_{t=0} \qquad (f \in \mathcal{S}(\mathbb{R}),\ x \in \mathbb{R}).$$

For $X = X_1 + iX_2 \in \mathfrak{g}_{\mathbb{C}}^J$ with $X_i \in \mathfrak{g}^J$, we set

$$d\pi_{SW}^m(X) = d\pi_{SW}^m(X_1) + i\, d\pi_{SW}^m(X_2).$$

3.2.1 Lemma. *The infinitesimal Schrödinger representation $d\pi_S^m$ acts on $\mathcal{S}(\mathbb{R})$ by the following operators:*

$$P = \frac{d}{dx} \qquad\qquad Y_+ = \frac{1}{2}\frac{d}{dx} - 2\pi m x$$

$$Q = 4\pi i m x \qquad\qquad Y_- = \frac{1}{2}\frac{d}{dx} + 2\pi m x$$

$$R = 2\pi i m \qquad\qquad Z_0 = 2\pi m$$

(The elements $P, Q, R, Y_{\pm}, Z_0$ are defined in Sections 1.3 resp. 1.4).

Proof: This is an easy exercise using the formula (2.1).   □

The infinitesimal Weil representation is more difficult to compute. First of all we need an explicit description of the cocycle λ defining the metaplectic group. For

$$M = \begin{pmatrix} a & b \\ c & d \end{pmatrix}, \qquad M' = \begin{pmatrix} a' & b' \\ c' & d' \end{pmatrix},$$

two elements of $SL(2,\mathbb{R})$, it is given by

$$\lambda(M, M') = (x(M), x(M'))(-x(M)x(M'), x(MM')), \tag{3.4}$$

where $(\ ,\ )$ denotes the Hilbert symbol and

$$x(M) = \begin{cases} c & \text{if } c \neq 0, \\ d & \text{if } c = 0. \end{cases}$$

(see [Ge2], p. 13–14). Define

$$\rho(t) = \begin{cases} -1 & \text{if } t > 0, \\ 1 & \text{if } t \leq 0. \end{cases}$$

Then, using the above description of λ, one can check that we have the following one-parameter subgroups $\mathbb{R} \to Mp(\mathbb{R})$ corresponding to the Lie algebra elements $F, G, H \in \mathfrak{g}^J$:

$$\phi_F(t) = \left(\begin{pmatrix} 1 & t \\ 0 & 1 \end{pmatrix}, 1 \right),$$

$$\phi_G(t) = \left(\begin{pmatrix} 1 & 0 \\ t & 1 \end{pmatrix}, \rho(t) \right),$$

$$\phi_H(t) = \left(\begin{pmatrix} e^t & 0 \\ 0 & e^{-t} \end{pmatrix}, 1 \right).$$

From this it is very easy to calculate $d\pi_W^m(F)$ and $d\pi_W^m(H)$, but there is a small difficulty in determining $d\pi_W^m(G)$. The best thing is to use the Fourier transformation $\mathcal{F}$. One computes

$$\mathcal{F} \circ d\pi_W^m(G) \circ \mathcal{F}^{-1} = -2\pi i m x^2,$$

and derives from this the formula in the following lemma.

3.2.2 Lemma. *We have the following formulas for the infinitesimal Weil representation $d\pi_W^m$ acting on $\mathcal{S}(\mathbb{R})$:*

$$F = 2\pi i m x^2 \qquad\qquad X_+ = \frac{1}{4} + \frac{1}{2}x\frac{d}{dx} - \pi m x^2 - \frac{1}{16\pi m}\frac{d^2}{dx^2}$$

$$G = \frac{i}{8\pi m}\frac{d^2}{dx^2} \qquad\qquad X_- = \frac{1}{4} + \frac{1}{2}x\frac{d}{dx} + \pi m x^2 + \frac{1}{16\pi m}\frac{d^2}{dx^2}$$

$$H = \frac{1}{2} + x\frac{d}{dx} \qquad\qquad Z = 2\pi m x^2 - \frac{1}{8\pi m}\frac{d^2}{dx^2}$$

(The elements $F, G, H, X_\pm, Z$ are defined in Sections 1.3 resp. 1.4).

We want to describe the infinitesimal Schrödinger-Weil representation in a purely algebraic way. Observe that π_{sw}^m, regarded as a representation of $\widetilde{G^J}$, decomposes over the maximal compact subgroup $\widetilde{K}$, which is a two-fold cover of $SO(2)$. Hence the element $F - G$, which spans the Lie algebra of this maximal compact subgroup, acts on an irreducible $\widetilde{K}$-module by ik, where $k \in \frac{1}{2}\mathbb{Z}$. In other words, $Z \in \mathfrak{g}_{\mathbb{C}}^J$ acts on irreducible $\widetilde{K}$-submodules of π_{sw}^m by half-integers. The subspace V of K-finite vectors therefore allows a decomposition

$$V = \sum_{k \in \frac{1}{2}\mathbb{Z}} V_k \quad \text{with} \quad V_k = \{v \in V : Zv = kv\}. \tag{3.5}$$

Moreover, every V_k is at most one-dimensional, because $Zv = kv$ is a second order differential equation for the Schwartz function v, and at most one of its solutions will lie in $\mathcal{S}(\mathbb{R})$. We further observe that π_{sw}^m is a **lowest (highest) weight representation** if $m > 0$ (resp. $m < 0$), i.e., there is a vector $v \in V$ such that $X_- v = Y_- v = 0$ (resp. $X_+ v = Y_+ v = 0$). The lowest (highest) weight vector is given by $e^{-2\pi|m|x^2}$. Now there is the following purely algebraic result.

3.2.3 Proposition. *Let $m \in \mathbb{R}^*$. There is exactly one lowest (resp. highest) weight representation of $\mathfrak{g}_{\mathbb{C}}^J$ on a space V which admits a decomposition (3.5) such that $\dim V_k \leq 1$ for all k, and such that Z_0 acts by $2\pi m$. In the lowest weight case, this representation has the space $V = \langle v_j \rangle_{j \in \mathbb{N}_0}$ and is given by*

$$Z_0 v_j = \mu v_j \qquad\qquad Z v_j = \left(j + \frac{1}{2}\right) v_j$$

$$Y_+ v_j = v_{j+1} \qquad\qquad X_+ v_j = -\frac{1}{2\mu} v_{j+2} \tag{3.6}$$

$$Y_- v_j = -\mu j v_{j-1} \qquad\qquad X_- v_j = \frac{\mu}{2} j(j-1) v_{j-2}$$

($v_{-1} = v_{-2} = 0$ understood). In the highest weight case, this representation has the space $V = \langle v_{-j} \rangle_{j \in \mathbb{N}_0}$ and acts by

$$Z_0 v_{-j} = \mu v_{-j} \qquad\qquad Z v_{-j} = -\left(j + \frac{1}{2}\right) v_{-j}$$

$$Y_- v_{-j} = v_{-(j+1)} \qquad\qquad X_- v_{-j} = \frac{1}{2\mu} v_{-(j+2)} \tag{3.7}$$

$$Y_+ v_{-j} = \mu j v_{-(j-1)} \qquad\qquad X_+ v_{-j} = -\frac{\mu}{2} j(j-1) v_{-(j-2)}$$

(with $v_1 = v_2 = 0$ understood).

3.2.4 Remark. In the lowest weight case, this representation may equivalently be characterized by

a) the existence of a lowest weight vector v_0 of weight 1/2, i.e. with

$$Z_0 v_0 = \mu v_0, \qquad Z v_0 = (1/2) v_0, \qquad Y_- v_0 = X_- v_0 = 0 \qquad \text{and}$$

b) the relation $D_+ = X_+ + \frac{1}{2\mu} Y_+^2 = 0$.

The **proof** of the above proposition (to be found in [Be1]) is straightforward: Starting with a vector v_0 of lowest weight k_0, one looks at

$$v_j := Y_+^j v_0 \qquad \text{and} \qquad v_{2j}' := X_+^j v_0,$$

and using the Lie algebra relations verifies, that v_{2j}' is a multiple of v_{2j} if and only if $2\mu X_+$ and Y_+^2 have the same action and if $k_0 = 1/2$ holds. □

From this proposition and the considerations before it, we see the following.

3.2.5 Corollary.

i) *If $m > 0$, the infinitesimal Schrödinger-Weil representation $d\pi_{SW}^m$ is given by the formulas (3.6), if in the space of K-finite vectors of $L^2(\mathbb{R})$ we set $v_0 = e^{-2\pi m x^2}$ and $v_{j+1} := Y_+ v_j$ for $j \geq 0$.*

ii) *If $m < 0$, then $d\pi_{SW}^m$ is given by the formulas (3.7), if in the space of K-finite vectors of $L^2(\mathbb{R})$ we set $v_0 = e^{2\pi m x^2}$ and $v_{-(j+1)} := Y_- v_{-j}$ for $j \geq 0$.*

3.2.6 Remark. Note that Proposition 3.2.3 actually yields more representations than the infinitesimal Schrödinger-Weil representations. But the additional ones do not come from unitary representations of the group.

The right half of the formulas (3.6) and (3.7) is nothing but the infinitesimal Weil representation, because π_W^m is just the restriction of π_{SW}^m to $\mathfrak{sl}_2$. We see that $\hat\pi_W^m$ decomposes into two irreducible components,

$$\hat\pi_W^m = \hat\pi_W^{m+} \oplus \hat\pi_W^{m-},$$

where $\hat\pi_W^{m+}$, the **positive** or **even Weil representation**, acts on the space spanned by the v_j with even indices, and $\hat\pi_W^{m-}$, the **negative** or **odd Weil representation**, acts on the space spanned by the v_j with odd indices (note that the sign in the symbol $\hat\pi_W^{m\pm}$ has nothing to do with the representation being of highest or lowest weight). The naming even and odd comes from the fact that if $\pi_W^{m\pm}$ is realized on the space of K-finite vectors in $L^2(\mathbb{R})$, then it consists entirely of even resp. odd functions. See also Proposition 2.5.1.

3.2.7 Corollary.

i) *Let $m > 0$. The infinitesimal even Weil representation $\hat\pi_W^{m+}$ acts on the space $\langle v_j \rangle_{j \in 2\mathbb{N}_0}$ by the right half of the formulas (3.6). Hence it is a lowest weight representation of lowest weight 1/2. The odd Weil representation $\hat\pi_W^{m-}$ acts on $\langle v_j \rangle_{j \in 2\mathbb{N}_0+1}$ and is a lowest weight representation of lowest weight 3/2.*

ii) Let $m < 0$. Then $\hat{\pi}_W^{m+}$ acts on the space $\langle v_{-j} \rangle_{j \in 2\mathbb{N}_0}$ by the right half of the formulas (3.7). It is a highest weight representation of highest weight $-1/2$. The odd Weil representation $\hat{\pi}_W^{m-}$ acts on $\langle v_{-j} \rangle_{j \in 2\mathbb{N}_0+1}$, and is a highest weight representation of highest weight $-3/2$.

From this corollary we immediately see how the even and odd Weil representations fit into the classification of metaplectic representations given in the preceding section:

3.2.8 Corollary.

i) For $m > 0$ we have

$$\hat{\pi}_W^{m+} = \hat{\pi}_{1/2}^+, \qquad \hat{\pi}_W^{m-} = \hat{\pi}_{3/2}^+.$$

ii) For $m < 0$ we have

$$\hat{\pi}_W^{m+} = \hat{\pi}_{1/2}^-, \qquad \hat{\pi}_W^{m-} = \hat{\pi}_{3/2}^-.$$

A model for principal and discrete series representations

The representations $(\hat{\pi}, V)$ enumerated in the last section may be realized by the action of the left invariant differential operators

$$\mathcal{L}_X \phi(g) = \frac{d}{dt} \phi(g \exp tX) \Big|_{t=0}$$

on functions ϕ living on $G^J(\mathbb{R})$. Here the S-coordinates $(x, y, \theta, p, q, \zeta)$ seem more appropriate than the EZ-coordinates. We use the notation

$$\phi(g) = \phi^S(x, y, \theta, p, q, \zeta).$$

The elements of the Lie algebra $\mathfrak{g}_{\mathbb{C}}^J$ may be viewed as the following differential operators on such functions ϕ:

$$
\begin{aligned}
\mathcal{L}_{Z_0} &= 2\pi \zeta \partial_\zeta, \\
\mathcal{L}_{Y_\pm} &= (1/2) y^{-1/2} e^{\pm i\theta} \Big(\partial_p - (x \mp iy) \partial_q - (p(x \mp iy) + q) 2\pi i \zeta \partial_\zeta \Big), \\
\mathcal{L}_{X_\pm} &= \pm (i/2) e^{\pm 2i\theta} \Big(2y(\partial_x \mp i\partial_y) - \partial_\theta \Big), \\
\mathcal{L}_Z &= -i\partial_\theta.
\end{aligned}
$$

We remind the reader that for these coordinates we have

$$x, p, q \in \mathbb{R}, \qquad y \in \mathbb{R}_{>0}, \qquad \theta \in \mathbb{R}/2\pi\mathbb{Z}, \qquad \zeta \in S^1.$$

In particular, to define a function ϕ on $G^J(\mathbb{R})$, ϕ^S has to be periodic in θ with period 2π. We will later on come up with functions ϕ^S with period 4π in θ.

These functions then will be thought of as functions on the metaplectic cover $\tilde{G}^J(\mathbb{R})$ of $G^J(\mathbb{R})$ (resp. on $\widetilde{SL}(2,\mathbb{R})$ if only x, y, θ appear). Having this in mind, we will often simply skip the suffix "S" and write $\phi(x, y, \theta, p, q, \zeta)$. Now, here is the first model.

3.2.9 Proposition. a) *The space* $V_{m,s,\nu} = \langle v_j \otimes w_l \rangle$ *for the principal series representation* $\hat{\pi}_{m,s,\nu}$ *is realized by*

$$v_j \otimes w_l = \phi_{m,s,j,l}, \qquad j \in \mathbb{N}_0, \ l \in 2\mathbb{Z} + \nu + 1/2$$

with

$$\phi_{m,s,j,l}(g) = \zeta^m e^{i(j+l)\theta} y^{(s+3/2)/2} e^m(pz)\psi_j(py^{1/2}).$$

b) *The space* $V_{m,k}^+ = \langle v_j \otimes w_l \rangle$ *for the discrete series representation* $\hat{\pi}_{m,k}^+$ *is realized by*

$$v_j \otimes w_l = c_l \phi_{m,k,j,l}^+ \qquad j \in \mathbb{N}_0, \ l \in 2\mathbb{N}_0$$

with

$$\phi_{m,k,j,l}^+(g) = \zeta^m e^{i(k+j+l)\theta} y^{k/2} e^m(pz)\psi_j(py^{1/2}),$$

$$c_l = (k-1/2)(k+1/2)\cdots(k-1/2+l+1).$$

c) *The space* $V_{m,k}^- = \langle v_j \otimes w_l \rangle$ *for the discrete series representation* $\hat{\pi}_{m,k}^-$ *is realized by*

$$v_j \otimes w_l = c_l \phi_{m,k,j,l}^- \qquad j \in \mathbb{N}_0, \ l \in -2\mathbb{N}_0$$

with

$$\phi_{m,k,j,l}^-(g) = \zeta^m e^{i(1-k+j+l)\theta} y^{k/2} e^m(pz)\psi_j(py^{1/2}),$$

$$c_l = (k-1/2)(k-3/2)\cdots(k-1/2-(l-1)).$$

In all cases ψ_j $(j \in \mathbb{N}_0)$ *is a family of polynomials in one variable, say* u, *with*

$$\psi_0(u) = 1, \quad \psi_{j+1} = (1/2)\psi_j' - 2\mu u\psi_j \quad \text{and} \quad \psi_j'' - 4\mu u\psi_j' + 4\mu j\psi_j = 0,$$

i.e., related to the Hermite polynomials $H_j(v)$ *by the substitution*

$$u = (2\mu)^{-1/2}v.$$

Proof: The functions given here arise from the construction of representations π of $G^J(\mathbb{R})$ by the induction process to be described below. Beside this, a direct computation shows that application of the differential operators $\mathcal{L}_X$ produces precisely the exact relations between the $v_j \otimes w_l$ required by Propositions 3.1.6 and 3.1.7 in the last section:

As is easily seen, the functions ϕ given in the proposition are products of functions $v_j = \tilde{\phi}_{m,1/2,j}$ with

$$\tilde{\phi}_{m,1/2,j}(\tilde{g}) = \zeta^m y^{1/4} e^{i(j+1/2)\theta} e^m(pz)\psi_j(py^{1/2}), \quad j \in \mathbb{N}_0,$$

and $w_l = \tilde{\Psi}_l$ with

$$\tilde{\Psi}_l(\tilde{g}) = y^{(s+1)/2} e^{i(l-1/2)\theta}, \qquad l \in 2\mathbb{Z} + \nu + 1/2, \quad \text{in case a)}$$

resp. $w_{\pm l} = c_{\pm l}\tilde{\Psi}_{\pm l}$ with

$$\tilde{\Psi}_{\pm l}(\tilde{g}) = y^{k/2-1/4} e^{\pm i(k-1/2+l)\theta}, \qquad l \in \mathbb{N}_0, \qquad \text{in cases b), c).}$$

Obviously the factors $\tilde{\phi}_{m,1/2,j}$ and $\tilde{\Psi}_l$ live on the metaplectic cover, but their product is a function on $G^J(\mathbb{R})$.

We have, for instance,

$$\mathcal{L}_{X_+}\tilde{\Psi}_l = (1/2)c_l(2k - 1 + l)\tilde{\Psi}_{l+2}$$

and by the relations prescribed by $\hat{\pi}_k^+$ this has to be

$$c_{l+2}\tilde{\Psi}_{l+2},$$

explaining the formula in the proposition given for the coefficients c_l, $l \in \mathbb{N}_0$. For $-l \in \mathbb{N}_0$ the computation goes the same way. In case a) there is no need of a "normalizing" constant on behalf of the symmetry of the relations for $\pi_{s,\nu}$ in the $+$ and $-$ direction.

Similarly, we get by application of the differential operator $\mathcal{L}_{X_+}$ to $v_j = \phi_{m,1/2,j}$

$$\mathcal{L}_{X_+}v_j = (1/2)(j + 1 - 4\mu p^2 y + py^{1/2}\psi_j'/\psi_j)e^{2i\theta}v_j.$$

By the relations for π_{sw}^m one has also

$$\mathcal{L}_{X_+}v_j = -(2\mu)^{-1}v_{j+2}.$$

Thus, we get

$$\psi_{j+2} = (4\mu^2 p^2 y - (j+1)/2)\psi_j - py^{1/2}\psi_j'.$$

And from

$$\mathcal{L}_{Y_+}v_j = (1/2)(2\mu py + (1/2)y^{1/2}\psi_j'/\psi_j)y^{-1/2}e^{i\theta}v_j = v_{j+1},$$

$$\mathcal{L}_{Y_+}v_j = (1/2)(\psi_j'/\psi_j)e^{-i\theta}v_j = -\mu j v_{j-1}$$

we deduce

$$\psi_{j+1} = (1/2)\psi_j' - 2\mu py^{1/2}\psi_j \quad \text{and} \quad \psi_j' = -2\mu j\psi_{j-1}.$$

With $u = py^{1/2}$ both equations combine to

$$\psi_j'' - 4\mu u \psi_j' + 4\mu j \psi_j = 0$$

and this is consistent with the equation above coming from $\mathcal{L}_{X_+} v_j$ and the corresponding expression for $\mathcal{L}_{X_-} v_j$ to be treated in the same way.

The unitarizability question

As in the general theory, we can decide here which of the given infinitesimal representations $\hat{\pi}$ listed in Theorem 3.1.9 may come from a unitary representation π of $G^J(\mathbb{R})$.

As, for instance, in [La] p. 122 one easily deduces that for a unitary (π, V) with scalar product $\langle \, , \, \rangle$ we have

$$\langle d\pi(X)v, \tilde{v} \rangle + \langle v, d\pi(X)\tilde{v} \rangle = 0 \quad \text{for all} \quad v, \tilde{v} \in V$$

and

$$d\pi(X_\pm) = -d\pi(X_\mp)^*, \quad d\pi(Y_\pm) = -d\pi(Y_\mp)^*,$$

if $X_\pm$ and $Y_\pm$ are the elements of $\mathfrak{g}_{\mathbb{C}}^J$ as above. Using this for the representation π_{SW}^m from Proposition 3.1.4 we see that for $V = \langle v_j \rangle_{j \in \mathbb{N}_0}$ to carry a scalar product $\langle \, , \, \rangle$ we necessarily have

$$\langle Y_+ v_j, v_{j+1} \rangle = -\langle v_j, Y_- v_{j+1} \rangle,$$

i.e., with $\mu = 2\pi m$,

$$\|v_{j+1}\|^2 = (j+1)\bar{\mu}\|v_j\|^2.$$

Thus we recover the following result, which may also be seen by inspection of the usual formulas for the Schrödinger representation as a representation on $L^2(\mathbb{R})$.

3.2.10 Remark. The Schrödinger-Weil representation $\hat{\pi}_{SW}^m$ is unitarizable.

In the same manner, we get for the principal series representation $\hat{\pi}_{s,\nu}$ of $\mathfrak{sl}_2$ from

$$\langle \hat{X}_+ w_l, w_{l+2} \rangle = -\langle w_l, \hat{X}_- w_{l+2} \rangle$$

the relation

$$(s + 1 + l - 1/2)\|w_{l+2}\|^2 = -(\bar{s} + 1 - (l + 2 - 1/2))\|w_l\|^2$$

i.e. for $l = 0$ and $\bar{s} \neq 1/2$ the condition

$$\frac{s + 1/2}{-\bar{s} + 1/2} > 0$$

which demands for

$$s \quad \text{real and} \quad s^2 < 1/4 \quad \text{or} \quad s \in i\mathbb{R}.$$

For $\hat{\pi}_{k_0}^+$ we come up with

$$\|w_{l+2}\|^2 = \frac{l+2}{2}\left(k_0 + \frac{l+2}{2} - 1\right)\|w_l\|^2$$

and for $\hat{\pi}_{k_0}^-$ with

$$\|w_{-(l+2)}\|^2 = \frac{l+2}{2}\left(k_0 + \frac{l+2}{2} - 1\right)\|w_{-l}\|^2$$

showing that in both cases we have to require $k_0 \geq 1/2$. All this put together gives the following result.

3.2.11 Proposition. *The representation $\hat{\pi}_{m,s,\nu}$ is unitarizable for $m > 0$ and $s \in i\mathbb{R}$ or $s \in \mathbb{R}$ with $s^2 < 1/4$, and $\hat{\pi}_{m,k}^\pm$ is unitarizable for $m > 0$ and $k \geq 1$.*

Proof: As $\hat{\pi}_{SW}^m$ is unitarizable by the last remark, a reasoning like in Proposition 5.9.1 below tells us that $\hat{\pi} = \tilde{\pi} \otimes \hat{\pi}_{SW}^m$ is unitarizable exactly if $\tilde{\pi}$ is. By the way, we can see this directly: As we have by Proposition 3.1.6

$$\hat{D}_+(v_j \otimes w_l) = (1/2)(s + 1 + (l - 1/2))v_j \otimes w_{l+2}$$

and

$$\hat{D}_-(v_j \otimes w_{l+2}) = (1/2)(s + 1 - (l + 2 - 1/2))v_j \otimes w_l,$$

we come up for $l = 0$ with the condition

$$\frac{s + 1/2}{-\bar{s} + 1/2} > 0,$$

exactly as above for the $\mathfrak{sl}_2$-case.

The scalar products defined for the generating elements $v_j \otimes w_l$ make the space spanned by these elements a pre-Hilbert space which may be completed to a Hilbert space. $\qquad\square$

It is to be remarked here, that $\hat{\pi}_{m,s,\nu}$, having infinite dimensional subspaces of fixed weight, is **not admissible.**

We can summarize and give the following **classification of irreducible unitary representations of $G^J(\mathbb{R})$:**

3.2.12 Theorem. *An irreducible unitary representation π of $G^J(\mathbb{R})$ with central character $e^{2\pi i m x}$, $m \in \mathbb{R}^*$, is infinitesimally equivalent to*

	a **continuous series representation** $\hat{\pi}_{m,s,\nu}$,	$s \in i\mathbb{R}$,
or	a **complementary series representation** $\hat{\pi}_{m,s,\nu}$,	$s \in \mathbb{R}$, $s^2 < \dfrac{1}{4}$,
or	a **positive discrete series representation** $\hat{\pi}_{m,k}^+$,	$k \geq 1$,
or	a **negative discrete series representation** $\hat{\pi}_{m,k}^-$,	$k \geq 1$.

The only equivalences between these representations are

$$\hat{\pi}_{m,s,\nu} \simeq \hat{\pi}_{m,-s,\nu},$$

all other representations are inequivalent.

Table 3.1 gives an overview over the irreducible, unitary representations of $G^J(\mathbb{R})$ with non-trivial central character $x \mapsto e^{2\pi i m x}$, $m \in \mathbb{R}^*$ (the statements about the Whittaker models $\mathcal{W}^{n,r}$ will be proved in Section 3.6 and are listed here for completeness).

name	principal series representation	pos. discrete series rep.	neg. discrete series rep.
symbol	$\hat{\pi}_{m,s,\nu}$	$\hat{\pi}^+_{m,k}$	$\hat{\pi}^-_{m,k}$
parameters	$s \in i\mathbb{R}$ (cont. ser.) or $s \in \mathbb{R}$, $s^2 < \frac{1}{4}$ (compl.)	$k \in \mathbb{Z}$, $k \geq 1$	$k \in \mathbb{Z}$, $k \geq 1$
isomorphic to	$\hat{\pi}_{s,\nu} \otimes \hat{\pi}^m_{SW}$	$\hat{\pi}^+_{k-1/2} \otimes \hat{\pi}^m_{SW}$	$\hat{\pi}^-_{k-1/2} \otimes \hat{\pi}^m_{SW}$
equivalences	$\hat{\pi}_{m,s,\nu} \simeq \hat{\pi}_{m,-s,\nu}$	none	none
admissible	no	iff $m > 0$	iff $m < 0$
$\mathcal{W}^{n,r}$ exists if	$mN > 0$	$mN > 0$	$mN < 0$
C operates by	$\frac{1}{2}(s^2 - 1)$	$\frac{1}{2}(k - \frac{1}{2})(k - \frac{5}{2})$	$\frac{1}{2}(k - \frac{1}{2})(k - \frac{5}{2})$

Table 3.1: Infinitesimal unitary representations of $\mathfrak{g}^J_{\mathbb{C}}$

3.3   Representations induced from B^J

In the SL(2)-theory, the representations of the principal and the discrete series may be realized as representations induced from the Borel subgroup NA resp., as subrepresentations of such. Now, the right way to carry this over to the Jacobi group is to induce from the group

$$B^J = AZN^J = \{b = \tilde{n}(x,q,\zeta)t(y) : x,q \in \mathbb{R}, \ y \in \mathbb{R}_{>0}, \ \zeta \in S^1\}$$

We denote by $\chi_{m,s}$ the character of B^J given by

$$\chi_{m,s}(\tilde{n}(x,q,\zeta)t(y)) := \zeta^m y^{s/2}, \tag{3.8}$$

which is unitary exactly for $s \in i\mathbb{R}$, and apply the machinery described in 2.1. Because of the commutation rule

$$t(y,p)\tilde{n}(x,q,\zeta) = \tilde{n}(x',q',\zeta')t(y,p)$$

with

$$x' = xy, \qquad q' = qy^{1/2} + pxy, \qquad \zeta' = \zeta e(2pqy^{1/2} + p^2xy),$$

it is easily seen that we have

$$d_r b = dx \, dq \frac{d\zeta}{\zeta} \frac{dy}{y} \qquad \text{and} \qquad \Delta_{B^J}(b) = y^{3/2}.$$

The decomposition

$$g = n(x)t(y)r(\theta)(p,q,\zeta) = \tilde{n}(x,\hat{q},\hat{\zeta})\tilde{t}(y)\hat{r}(\theta,\hat{p})$$

with

$$\hat{p} = py^{1/2}, \qquad \hat{q} = a + px, \qquad \hat{\zeta} = \zeta e(p(px+q)),$$

shows that the Borel section s used to construct the induced representation may be chosen here to be

$$\mathcal{K} = B^J \backslash G^J \ni (\theta,\hat{p}) \mapsto r(\theta,\hat{p}) \in G^J$$

such that we come out with the quasi-invariant measure μ_s given by

$$d\mu_s = d\hat{p}\,d\theta = y^{1/2}dp\,d\theta. \tag{3.9}$$

Now the prescription given in 2.1 produces the induced representation

$$\pi_{m,s} := \text{ind}_{B^J}^{G^J} \chi_{m,s}$$

given by right translation ρ on the space $\mathcal{H}_{m,s}$ of measurable functions ϕ on G^J with

i) $\qquad \phi(b_0 g) = y_0^{(s+3/2)/2}\zeta_0^m \phi(g)$

ii) $\qquad \|\phi\|_{\mathcal{H}_{m,s}}^2 = \int_{\mathcal{K}} |\phi(r(\theta,\hat{p})|^2 d\theta\,d\hat{p} < \infty.$

$\mathcal{H}_{m,s}$ is a Hilbert space with the scalar product.

$$\langle \phi_1, \phi_2 \rangle = \int_{\mathcal{K}} \phi_1(r(\theta,\hat{p}))\overline{\phi_2(r(\theta,\hat{p}))}d\theta\,d\hat{p} \tag{3.10}$$

The decomposition above shows that these functions ϕ are of the type

i') $\qquad \phi(g) = y^{(s+3/2)}\zeta^m e^m(p(px+q))\varphi(\theta,py^{1/2})$

ii') $\qquad \int |\varphi(\theta,v)|^2 \, d\theta\,dv < \infty.$

Remembering that the space of functions φ with (ii') can be spanned by functions of the type

$$\varphi(\theta, v) = e^{il\theta} e^{-v^2} H_j(v), \qquad l \in \mathbb{Z}, \quad j \in \mathbb{N}_0,$$

with the Hermite polynomials $H_j(v)$ we obtain after the substitution

$$v = (2\pi m)^{1/2} u, \qquad H_j((2\pi m)^{1/2} u) =: \psi_j,$$

the following statement.

3.3.1 Proposition. *For each integer $m > 0$ and each $s \in \mathbb{C}$ there is a representation $\pi_{m,s}$ of $G^J(\mathbb{R})$ given by right translation on the Hilbert space $\mathcal{H}_{m,s}$ spanned by the family of functions*

$$\phi'_{m,s,j,l}(g) = \zeta^m y^{(s+3/2)/2} e^{il\theta} e^m(pz) \psi_j(py^{1/2}), \qquad l \in \mathbb{Z}, \quad j \in \mathbb{N}_0.$$

By the general theory $\pi_{m,s} = \mathrm{ind}_{B^J}^{G^J} \chi_{m,s}$ is unitary if $\chi_{m,s}$ is, i.e., for $s \in i\mathbb{R}$. The question of irreducibility will be answered by the following comparison with the infinitesimal results. As the representation of G^J on $\mathcal{H}_{m,s}$ is given by right translation

$$\pi_{m,s}(g_0)\phi(g) = \phi(gg_0),$$

its derived representation $d\pi_{m,s}$ is given by

$$d\pi_{m,s}(X)\phi(g) = \frac{d}{dt}\phi(g \exp t X)\Big|_{t=0}$$

i.e., produces the operators realizing $\mathfrak{g}^J$ resp. $\mathfrak{g}_{\mathbb{C}}^J$.

As we can identify the functions spanning the representation spaces in Proposition 3.2.9 with those appearing here by

$$
\begin{array}{llll}
\phi_{m,s,j,l} & = & \phi'_{m,s,j,j+l} & , \quad l \in 2\mathbb{Z} \pm 1 \quad \text{for} \quad \hat{\pi}_{m,s,\nu} \\[2mm]
\phi^+_{m,k,j,l} & = & \phi'_{m,k-3/2,j,k+j+l} & , \quad l \in 2\mathbb{N}_0 \quad \text{for} \quad \hat{\pi}^+_{m,k} \\[2mm]
\phi^-_{m,1-k,j,l} & = & \phi'_{m,k-3/2,j,1-k+j+l} & , \quad l \in 2\mathbb{N}_0 \quad \text{for} \quad \hat{\pi}^-_{m,k}
\end{array}
$$

we may state:

3.3.2 Corollary. *The family $\{\phi'\}$ of functions above span invariant subspaces of $\mathcal{H}_{m,s}$ resp. $\mathcal{H}_{m,k-3/2}$ and thus give models for the representations $\pi_{m,s,\nu}$ resp. $\pi^\pm_{m,k}$ by right translations.*

3.3.3 Remark. It is perhaps not without interest that by the same reasoning the Schrödinger-Weil representation π^m_{SW} is realized for $m \in \mathbb{N}$ by the functions $\phi_{m,-1,j,-(j+1/2)}$, $j \in \mathbb{N}_0$ living on the metaplectic cover $\tilde{G}^J(\mathbb{R})$. And for the "mirror image" π^m_{SW} with $m < 0$ we come up with a space spanned by the family of functions $\bar{\phi}_{m,-1,j,j+1/2}$, $j \in \mathbb{N}_0$, complex conjugate to those above, and with $-m \in \mathbb{N}$.

3.4 Representations induced from K^J and the automorphic factor

We have the Iwasawa decomposition

$$G^J(\mathbb{R}) = N^J(\mathbb{R})A^J(\mathbb{R})K^J(\mathbb{R}),$$

the projection

$$pr: \ G^J(\mathbb{R}) \longrightarrow G^J(\mathbb{R})/K^J(\mathbb{R}) = \mathbf{X}$$

with $\mathbf{x}_0 = (i,0) \in \mathbf{X}$ and $\tau = x + iy$, $z = p\tau + q$, given by

$$g = ((u(x)t(y)r(\theta)), \ (p,q,\kappa)) \mapsto g(\mathbf{x}_0) = (\tau, z),$$

and the natural section

$$s: \ \mathbf{X} \longrightarrow G^J(\mathbb{R}),$$

for $\mathbf{x} = (\tau, z) = (x+iy, p\tau + q)$ given by

$$s(\mathbf{x}) = (p, q, 0)n(x)t(y) =: \ g_\mathbf{x}.$$

Then it is a very natural thing to take the representation $\chi^*_{m,k}$ of

$$K^J(\mathbb{R}) = \{r(\theta, \kappa) = r(\theta)\kappa : \ \theta, \ \kappa \in \mathbb{R}\}$$

for $m, k \in \mathbb{Z}$ given by

$$\chi^*_{m,k}(r(\theta, \kappa)) = e^{ik\theta}e^m(\kappa) \tag{3.11}$$

and to induce from here to $G^J(\mathbb{R})$, i.e., to look at the representation

$$\pi^*_{m,k} := \mathrm{ind}_{K^J(\mathbb{R})}^{G^J(\mathbb{R})} \chi^*_{m,k}$$

given by left translation

$$\lambda(g_0)\phi(g) \ = \ \phi(g_0^{-1}g)$$

on the space $B_{\chi^*_{m,k}}$ of measurable functions ϕ on $G^J(\mathbb{R})$ with

$$\phi(gr(\theta, \kappa)) = \phi(g)\chi^*_{m,k}(r(\theta, \kappa)) \qquad \text{for all } g \in G^J, \ r(\theta, \kappa) \in K^J,$$

and

$$\int |\phi((p,q,0)n(x)t(y))|^2 \, d\mu_\mathbf{x} < \infty$$

for $d\mu_\mathbf{x} = y^{-2} \, dx \, dy \, dp \, dq$.

There is an equivalent model for this representation given by

$$\pi^*(g_0)F(\mathbf{x}) = F(g_0^{-1}(\mathbf{x}))e^{ik\theta_0}e^m(\kappa_0) \quad \text{for} \quad F \in L^2(\mathbf{X}, d\mu_{\mathbf{x}}).$$

These considerations may be refined considerably by the following standard procedure leading to the **canonical automorphic factors**, which is a special case of a general discussion by Satake ([Sa1] or [Sa2] pp. 118-119) on the basis of older work (see for instance Matsushima and Murakami [MM] and Borel and Baily [BB] 1.9). As already remarked in Section 1.4, $G^J(\mathbb{R})$ is a group of Harish–Chandra type. More precisely, we have the following situation. The complexified Lie algebra $\mathfrak{g}_{\mathbb{C}}^J$ of G^J is given by

$$\mathfrak{g}_{\mathbb{C}}^J = \mathfrak{p}_0 \oplus \mathfrak{p}_+ \oplus \mathfrak{p}_- \qquad \text{with} \quad \mathfrak{p}_\pm = \langle X_\pm, Y_\pm \rangle_{\mathbb{C}}, \quad \mathfrak{p}_0 = \langle Z, Z_0 \rangle_{\mathbb{C}}$$

and (see 1.4)

$$[\mathfrak{p}_0, \mathfrak{p}_\pm] \subset \mathfrak{p}_\pm, \quad \overline{\mathfrak{p}_+} = \mathfrak{p}_-,$$

i.e. Satake's condition (HC 1). And for the complex groups

$$P_\pm = \exp \mathfrak{p}_\pm, \quad P_0 = \exp \mathfrak{p}_0,$$

we have Satake's condition (HC 2), i.e.

$$G^J \hookrightarrow P_+ P_0 P_- \hookrightarrow G_{\mathbb{C}}^J = \mathrm{SL}(2,\mathbb{C}) \ltimes H(\mathbb{C}). \tag{3.12}$$

This will become more transparent if we twist by a "**partial Cayley transform**" $\tilde{C}$, which in the symplectic part gives the usual bounded realization of the upper half plane $\mathbf{H}$ by the unit disc $\mathcal{D}$: We take the matrices

$$C = \begin{pmatrix} i & i \\ -1 & 1 \end{pmatrix} \quad \text{and} \quad C^{-1} = \frac{1}{2i}\begin{pmatrix} 1 & -i \\ 1 & i \end{pmatrix}$$

which induce a biholomorphic map

$$\mathbf{H} \xrightarrow{C^{-1}} \mathcal{D}, \tag{3.13}$$

$$\tau \longmapsto \tau^* = C^{-1}(\tau) = \frac{\tau - i}{\tau + i},$$

and twist $G = \mathrm{SL}(2, \mathbb{R})$ into

$$\begin{aligned} G^* &= C^{-1}GC = \mathrm{SU}(1,1) \\ &= \left\{ M^* = \begin{pmatrix} \alpha & \beta \\ \bar{\beta} & \bar{\alpha} \end{pmatrix} : \alpha, \beta \in \mathbb{C}, |\alpha|^2 - |\beta|^2 = 1 \right\} \end{aligned}$$

with

$$2\alpha = a + d + (b - c)i, \qquad 2\beta = a - d - (b + c)i \qquad \text{for} \quad M = \begin{pmatrix} a & b \\ c & d \end{pmatrix}.$$

Moreover $G^J(\mathbb{R}) = \mathrm{SL}(2,\mathbb{R}) \ltimes H(\mathbb{R})$ is twisted by $\tilde{C}$ into

$$G^{*J}(\mathbb{R}) = \tilde{C}^{-1} G^J(\mathbb{R}) \tilde{C} = \left\{ g^* = \begin{pmatrix} \alpha & \beta \\ \bar{\beta} & \bar{\alpha} \end{pmatrix} (X^*, \kappa) \right\}$$

with

$$X^* = X\tilde{C} = (\lambda i - \mu, \ \lambda i + \mu) \qquad \text{for } X = (\lambda, \mu).$$

In particular, $K^J = \mathrm{SO}(2) \times \mathbb{R} = \{ r(\theta, \kappa) : \theta, \ \kappa \in \mathbb{R} \}$ is twisted into

$$K^{*J} = \left\{ \begin{pmatrix} \gamma^{-1} & 0 \\ 0 & \gamma \end{pmatrix} (0, 0, \kappa) = d(\gamma^{-1}, \kappa) : \gamma \in S^1, \ \kappa \in \mathbb{R} \right\} \tag{3.14}$$

with

$$\gamma = e^{-i\theta}.$$

And we get a biholomorphic map $\tilde{C}^{-1} : \mathbf{H} \times \mathbb{C} \to \mathcal{D} \times \mathbb{C}$ given via the commutativity of the diagram

$$
\begin{array}{ccc}
G^J & \longrightarrow & G^{*J} \\
\downarrow & & \downarrow \\
\mathbf{H} \times \mathbb{C} = G^J/K^J & \longrightarrow & G^{*J}/K^{*J} = \mathcal{D} \times \mathbb{C} = \mathcal{D}^J \\
(\tau, z) & \longmapsto & (\tau^*, z^*) = \left(\dfrac{\tau - i}{\tau + i}, \ \dfrac{z}{\tau + i} \right).
\end{array}
$$

We will use the notation

$$N_+^J = \left\{ \begin{pmatrix} 1 & \tau^* \\ 0 & 1 \end{pmatrix} (0, z^*, 0) = n(\tau^*, z^*) : \tau^*, z^* \in \mathbb{C} \right\} \tag{3.15}$$

and

$$N_-^J = \left\{ \begin{pmatrix} 1 & 0 \\ u & 1 \end{pmatrix} (v, 0, 0) = n_-(u, v) : u, v \in \mathbb{C} \right\}. \tag{3.16}$$

Then we have the following central fact.

3.4.1 Remark. Each

$$g^* = M^*(Y^*, \kappa^*) \in G^{*J}$$

may be uniquely decomposed as

$$g^* = g_+^* g_0^* g_-^*$$

with

$$
\begin{aligned}
g_-^* &= n_-(u, v), \ u = c^*/d^*, \ v = \lambda^* - \mu^* c^*/d^* \\
g_0^* &= d(\gamma^{-1}, \kappa), \ \gamma = d^*, \ \kappa = \kappa^* + \mu^*(\lambda^* - \mu^* c^*/d^*) \\
g_+^* &= n(\tau^*, z^*), \ \tau^* = b^*/d^*, \ z^* = \mu^*/d^*.
\end{aligned}
$$

Since for $M^+ \in SU(1,1)$ we have $a^* = \beta$ and $d^* = \bar{\alpha}$ with $|\alpha|^2 - |\beta|^2 = 1$, here is

$$|\tau^*| = |\beta/\bar{\alpha}| < 1.$$

From this remark it may be concluded that

$$\begin{aligned} G^{*J}/K^{*J} &\xrightarrow{\sim} \mathcal{D} \times \mathbb{C} =: \mathcal{D}^J, \\ g^* &\longmapsto (\tau^*, z^*), \end{aligned}$$

is a **Harish–Chandra realization** of $\mathbf{H} \times \mathbb{C}$ as a "partially bounded domain". For the operation of G^* on $\mathcal{D}^J$ we have

3.4.2 Remark. An element $g^* = M^*(X^*, \kappa^*)$ acts on $(\tau^*, z^*) \in \mathcal{D}$ by

$$g^*(\tau^*, z^*) = \left(M^*(\tau^*), \frac{z^* + \lambda^* \tau^* + \mu^*}{c^* \tau^* + d^*} \right). \tag{3.17}$$

This easily comes out by the decomposition

$$g^* n(\tau^*, z^*) = (g^* n)_+ (g^* n)_0 (g^* n)_-.$$

We get

$$(g^* n)_+ = n(\tilde{\tau}^*, \tilde{z}^*)$$

with

$$\tilde{\tau}^* = M^*(\tau^*) = \frac{a^* \tau^* + b^*}{c^* \tau^* + d^*}, \quad \tilde{z}^* = \frac{z^* + \lambda^* \tau^* + \mu^*}{c^* \tau^* + d^*}$$

and

$$(g'n)_0 = d(\gamma^{-1}, \kappa),$$

where

$$\begin{aligned} \gamma &= (c^* \tau^* + d^*), \\ \kappa &= \kappa^* - c^* \frac{(z^* + \lambda^* \tau^* + \mu^*)^2}{c^* \tau^* + \mu^*} + \lambda^{*2} \tau^* + 2\lambda^* z^* + \lambda^* \mu^*. \end{aligned}$$

The element $(g^* n)_0 = d(\gamma^{-1}, \kappa) \in K_{\mathbb{C}}^{*J}$ is called the **canonical automorphic factor** and denoted by $J^*(g^*, (\tau^*, z^*))$. Its significance becomes clear if we define a unitary representation of G^{*J} by a method very close to the induction procedure described in 2.1, and which, as in the usual case (see for instance Knapp [Kn] p. 150), establishes the "holomorphic discrete series":

We denote by B_-^J the complex group

$$B_-^J = \left\{ b = \begin{pmatrix} a & 0 \\ c & a^{-1} \end{pmatrix} (\lambda, 0, \kappa) : a \in \mathbb{C}^*, \ c, \lambda, \kappa \in \mathbb{C} \right\}, \tag{3.18}$$

and for $m, k \in \mathbb{N}_0$ by $\xi_{m,k}$ its character given by

$$\xi_{m,k}(b) = a^k e^m(\kappa). \tag{3.19}$$

As we have

$$G^{*J}(\mathbb{R}) \hookrightarrow G^{*J}(\mathbb{R})B_-^J \simeq \mathcal{D}^J K_c^J N_-^J \hookrightarrow G^J(\mathbb{C})$$

with a complex structure on the term in the middle, we get

3.4.3 Proposition. *A unitary representation $\pi_{m,k}$ of $G^{*J}(\mathbb{R})$ is given by left-translation on the space $V_{\xi_{m,k}}$ of holomorphic functions ϕ living on $G^{*J}B_-^J$ with the properties*

i) $\qquad \phi(gb) = \xi_{m,k}(b)\phi(b) \qquad\qquad$ *for all $g \in G^{*J}B_-^J$ and $b \in B_-^J$,*

ii) $\qquad \|\phi\|^2 := \displaystyle\int_{G^{*J}(\mathbb{R})/Z} |\phi(g^*)|^2 \, dg^* < \infty,$

where dg is a Haar measure which will be normalized a bit further down.

It is not evident but true (and can be proved by the usual method as for instance in [La] pp. 182-184) that $V_{\xi_{m,k}}$ is a Hilbert space and that this representation is irreducible.

As in the usual procedure described in 2.1, there is a **second realization** of this representation given by restriction of the functions ϕ to $\mathcal{D}^J$. Writing

$$\phi(g^*) = \phi(\tau^*, z^*, \gamma, \kappa, u, v) \text{ for } g^* = n(\tau^*, z^*)d(\gamma^{-1}, \kappa)n(u, v),$$

we put

$$F(\tau^*, z^*) := \phi(\tau^*, z^*, 1, 0, 0, 0)$$

and get a holomorphic function F defined on $\mathcal{D}^J$. Obviously the map

$$r : \phi \mapsto F$$

from $V_{\xi_{m,k}}$ to holomorphic functions on $\mathcal{D}^J$ is for g^* given as above inverted by

$$r^* : F \longmapsto \phi_F, \qquad \phi_F(g) = \gamma^{-k} e^m(\kappa) F(\tau^*, z^*).$$

r is an interwining operator if the action of $g_0^* \in G^{*J}$ on F is defined by

$$(\pi_{m,k}^*(g_0^{-1})F)(\tau^*, z^*) = j_{k,m}(g_0^*, (\tau^*, z^*))F(g_0^*(\tau^*, z^*)),$$

where

$$j_{k,m}(g^*, (\tau^*, z^*)) = \xi_{m,k}(J^*(g^*, (\tau^*, z^*)))$$

$$= (c^*\tau^* + d^*)^{-k} e^m \left(\kappa^* - c^* \frac{(z^* + \lambda^*\tau^* + \mu^*)^2}{c^*\tau^* + d^*} + \lambda^{*2}\tau^* + 2\lambda^*z^* + \lambda^*\mu^* \right)$$

r is unitary if we define the norm of F as follows. We have as a G^{*J}-invariant measure on $\mathcal{D}^J$

$$d\mu_{\mathcal{D}} = (1 - (x^{*2} + y^{*2}))^{-3} \, dx^* \, dy^* \, d\xi^* \, dy^*$$

for $\tau^* = x^* + iy^*$, $\tau^* = \xi^* + iy^*$ (as can be seen for instance by pullback from the G^J-invariant measure on $\mathbf{H} \times \mathbb{C}$). We normalize such that we have

$$dg^* = d\mu_{\mathcal{D}^J} \, dk \quad \text{with} \quad \int_{K^{*J}/\mathbb{Z}} dk = 1. \tag{3.20}$$

The decomposition

$$G^{*J} \ni g^* = g^*_{(\tau^*, z^*)} k^*, \qquad k^* \in K^{*J}, \ g^*_{(\tau^*, z^*)}(0, 0) = (\tau^*, z^*),$$

and the unitarity of $\xi_{m,k}$ allow to write

$$
\begin{aligned}
\|\phi_F\|^2 &= \int_{G^{*J}/\mathbb{Z}} |\phi_\beta(g^*)|^2 \, dg \\
&= \int_{G^{*J}/\mathbb{Z}} |F(g^*(0,0))|^2 |j_{k,m}(g, (0,0))|^2 \, dg \\
&= \int_{\mathcal{D}^J} |F(\tau^*, z^*)|^2 |j_{k,m}(g_{(\tau^*, z^*)}, (0,0))|^2 \, d\mu_{\mathcal{D}^J}.
\end{aligned}
$$

This expression will be taken as $\|F\|^2$, and a small computation shows that we may also write

$$\|F\|^2 = \int_{\mathcal{D}^J} |F(\tau^*, z^*)|^2 e^{-m} \left(\frac{\bar{\tau}^* z^{*2} - \tau^* \bar{z}^{*2}}{1 - |\tau^*|^2} \right) (1 - |\tau^*|^2)^{k-3} \, dx^* \, dy^* \, d\xi^* \, dy^*.$$

To summarize, we get another model for the representation of the last proposition.

3.4.4 Corollary. *We get another realization of the representation $\pi_{m,k}$ of $G^{*J}(\mathbb{R})$ by the action*

$$(\pi_{m,k}(g_0^{*-1})F)(\tau^*, z^*) = j^*_{k,m}(g_0^*, (\tau^*, z^*)) F(g_0^*(\tau^*, z^*))$$

with

$$
\begin{aligned}
& j^*_{k,m}(g^*, (\tau^*, z^*)) \\
& = (c^*\tau^* + d^*)^{-k} e^m \left(\kappa^* - c \frac{(z^* + \lambda^*\tau^* + d^*)^2}{c^*\tau^* + d^*} + \lambda^{*2}\tau^* + 2\lambda^* z^* + \lambda^*\mu^* \right)
\end{aligned}
$$

for holomorphic functions F on $\mathcal{D}^J = \mathcal{D} \times \mathbb{C}$ with $\|F\| < \infty$.

Now, via

$$G^J \ni g \longmapsto \tilde{C}^{-1}g\tilde{C} = g^* \in G^{*J}$$

the representations of G^{*J} may be "untwisted" to get representations of G^J. We will do this and at the same time go back by the Cayley transformation from $\mathcal{D}^J$ to the biholomorphic equivalent space

$$\mathbf{X} = \mathbf{H} \times \mathbb{C} = G^J(\mathbb{R})/K^J(\mathbb{R}).$$

Denoting as above

$$d\mu_{\mathcal{D}^J,m,k} = |j_{k,m}(g^*_{(\tau^*,z^*)}, (0,0))|^2 \, d\mu_{\mathcal{D}^J}$$

and $L^2_{hol}(\mathcal{D}^J, d\mu_{\mathcal{D}^J,m,k})$ for the space of square integrable holomorphic functions F on $\mathcal{D}^J$, we associate to F a function $e^{-1}_{m,k}F$ on $\mathbf{X}$ given by

$$(C^{-1}_{m,k}F)(\tau, z) = F(\tilde{C}^{-1}(\tau, z))j_{k,m}(\tilde{C}, \tilde{C}^{-1}(\tau, z))^{-1}. \tag{3.21}$$

This map is inverted by

$$(C_{m,k}f)(\tau, z^*) = f(\tilde{C}(\tau, z^*))j_{k,m}(\tilde{C}, (\tau^*, z^*)).$$

Here we have used the fact that the definition of the automorphic factor

$$G^{*J} \times \mathcal{D}^J \longrightarrow \mathbb{C}$$

given above also works for any pair $(g^*, (\tau^*, z^*))$, $g^* \in G^J(\mathbb{C})$, $(\tau^*, z^*) \in \mathbb{C}^2$ such that $g^*g^*_{(\tau^*,z^*)} \in K^{*J}_c$. Now, by the maps $C_{m,k}$ and $C^{-1}_{m,k}$ the representation $\pi^*_{m,k}$, for $g^* \in G^{*J}$ given by

$$\pi^*_{m,k}(g^{*-1})F((\tau^*, z^*)) = j_{k,m}(g^*, (\tau^*, z^*))F(g, (\tau^*, z^*)),$$

is intertwined to a representation $\pi_{m,k}$ of G^J given by

$$\pi_{m,k}(g^{-1})f(\tau, z) = j_{k,m}(g, (\tau, z))f(g(\tau, z))$$

with

$$j_{k,m}(g, (\tau, z)) = j_{k,m}(\tilde{C}^{-1}g\tilde{C}, \ \tilde{C}^{-1}(\tau, z))$$

$$\cdot j_{k,m}(\tilde{C}, \tilde{C}^{-1}g(\tau, z))j_{k,m}(C, C^{-1}(\tau, z))^{-1}.$$

Here the function

$$j_{k,m} : G^J \times \mathbf{X} \longrightarrow \mathbb{C}$$

is defined by the same prescription as $j : G^* \times \mathcal{D}^J \to \mathbb{C}$ which comes out if one uses the cocycle condition of the automorphic factor to reduce the right hand side.

$C_{m,k}$ maps $L^2_{hol}(\mathcal{D}^J, d\mu_{\mathcal{D}^J,m,k})$ isomorphically into the space $L^2_{hol}(\mathbf{X}, d\mu_{\mathbf{X},m,k})$ of holomorphic functions on $\mathbf{X} = \mathbf{H} \times \mathbb{C}$, square integrable with respect to

$$
\begin{aligned}
d\mu_{\mathbf{X},m,k} &= |j_{k,m}(g_{(\tau,z)},(i,0))|^2 \, d\mu_{\mathbf{X}} \\
&= e^{-4\pi m \eta^2/y} y^{k-3} \, dx \, dy \, d\xi \, d\eta.
\end{aligned}
$$

Here we use again a decomposition

$$
G^J(\mathbb{R}) \ni g = g_{(\tau,z)} k
$$

where $k \in K^J$ and

$$
g_{(\tau,z)} = \left[\begin{pmatrix} y^{1/2} & xy^{-1/2} \\ 0 & y^{-1/2} \end{pmatrix}, \ (py^{1/2}, pxy^{-1/2} + qy^{-1/2}, 0) \right]
$$

is such that $g_{(\tau,z)}(i,0) = (\tau, z)$. Summing up, we have the

3.4.5 Proposition. *A unitary representation $\pi_{m,k}$ of $G^J(\mathbb{R})$ is given by*

$$
\pi_{m,k}(g^{-1})f(\tau,z) = f(g,(\tau,z))j_{k,m}(g,(\tau,z)) \quad \text{for} \quad f \in L^2_{hol}(\mathbf{X}, d\mu_{\mathbf{X},m,k}).
$$

By the usual lifting

$$
f \longmapsto \phi_f \quad \text{with} \quad \phi_f(g) = f(g(i,0))j_{k,m}(g,(i,0))
$$

we get a model of this representation by functions ϕ living on $G^J(\mathbb{R})$:

3.4.6 Corollary. *A unitary representation $\pi_{m,k}$ of $G^J(\mathbb{R})$ is given by*

$$
\pi_{m,k}(g_0^{-1})\phi(g) = \phi(g_0 g), \qquad g_0, g \in G^J(\mathbb{R}),
$$

for functions ϕ of the type

$$
\phi(g) = e^m(k)e^{ik\theta}y^{k/2}e^m(pz)f(g(i,0)),
$$

where here $g = (M, p, q, \kappa) \in G^J$ in the S-coordinates and

$$
f \in L^2_{hol}(\mathbf{H} \times \mathbb{C}, \ d\mu_{\mathbf{H} \times \mathbb{C},m,k}),
$$

or equivalently f is of the type

$$
f(\tau,z) = F\left(\frac{\tau - i}{\tau + i}, \frac{z}{\tau + i}\right)\left(\frac{\tau + i}{2i}\right)^k e^m\left(-\frac{z}{\tau + i}\right)
$$

with $F \in L^2_{hol}(\mathcal{D}^J, d\mu_{\mathcal{D}^J,m,k})$.

Obviously, we have arrived at a subrepresentation of the representation

$$
\mathrm{ind}_{K^J(\mathbb{R})}^{G^J(\mathbb{R})} \xi^*_{m,k}
$$

from the beginning of this section.

3.5 Differential operators on $\mathbf{X} = \mathbf{H} \times \mathbb{C}$

After having provided an "explanation" of the canonical automorphic factor $j_{k,m}(g,(\tau,z))$ for $G^J(\mathbb{R})$ we will interrupt for a moment the discussion of the models for the representations π of $G^J(\mathbb{R})$ and study the algebras of in- and covariant differential operators on $\mathbf{X} = \mathbf{H} \times \mathbb{C}$.

We use again the S-coordinates

$$(\tau = x + iy,\ z = p\tau + q) \in \mathbf{H} \times \mathbb{C},$$

and by the usual misuse of notation write

$$f(x,y,p,q) = f(\tau,\bar{\tau},z,\bar{z})$$

for infinitely differentiable $\mathbb{C}$–valued functions f living on $\mathbf{H} \times \mathbb{C}$. We abbreviate

$$j_{k,m}(g) := j_{k,m}(g,(i,0)) = e^m(\kappa)y^{k/2}e^{ik\theta}e^m(pz).$$

3.5.1 Remark. Applying the $G^J(\mathbb{R})$-left invariant differential operators $\mathcal{L}_X$, $X \in \mathfrak{g}_{\mathbb{C}}^J$ from 1.4 to the functions

$$\phi = j_{k,m}f$$

living on G^J we get after a small calculation

$$\mathcal{L}_{Z_0}\phi = j_{k,m}2\pi m f, \qquad\qquad \mathcal{L}_Z\phi = j_{k,m}kf,$$
$$\mathcal{L}_{Y_\pm}\phi = j_{k\pm1,m}Y_\pm^{m,k}f, \qquad\qquad \mathcal{L}_{X_\pm}\phi = j_{k\pm2,m}X_\pm^{m,k}f,$$

with

$$
\begin{aligned}
Y_+f = Y_+^{m,k}f &= (1/2y)(f_p - (x-iy)f_q - 4py2\pi m\, f)\\
&= i\left(f_z + \frac{z-\bar{z}}{\tau-\bar{\tau}}4\pi im\, f\right)\\
Y_-f = Y_-^{m,k}f &= (1/2)(f_p - (x+iy)f_q) = (1/2)(\tau-\bar{\tau})f_{\bar{z}}\\
X_+f = X_+^{m,k}f &= i(f_x - if_y) + (4\pi imp^2 - ik/y)f\\
&= i\left(2f_\tau + 2\frac{z-\bar{z}}{\tau-\bar{\tau}}f_z + \left(4\pi im\left(\frac{z-\bar{z}}{\tau-\bar{\tau}}\right)^2 + \frac{2k}{\tau-\bar{\tau}}\right)f\right)\\
X_-f = X_-^{m,k}f &= -iy^2(f_x + if_y)\\
&= (i/2)(\tau-\bar{\tau})((\tau-\bar{\tau})f_{\bar{\tau}} + (z-\bar{z})f_{\bar{z}}).
\end{aligned}
$$

We see immediately that the holomorphy of f is equivalent to

$$Y_-f = X_-f = 0. \tag{3.22}$$

The operator $D_+ = X_+ + (4\pi m)^{-1} Y_+^2$ from the discussion of the representations of $\mathfrak{g}_{\mathbb{C}}^J$ in 3.1 induces the differential operator

$$D_+ = X_+ + (4\pi m)^{-1} Y_+^2 = \partial_\tau - (4\pi m)^{-1} \partial_z^2 + (2k-1)(\tau - \bar{\tau})^{-1}$$

which for $k = 1/2$ may be recognized as the heat operator. It was used in [Be2] for a characterization of the Jacobi theta function.

We use again the symbol $\pi_{m,k}$ to indicate the action of G^J on $C^\infty(\mathbf{X})$ given by

$$\pi_{m,k}(g^{-1}) f(\tau, z) = (f|_{k,m} g)(\tau, z) = j_{k,m}(g, (\tau, z)) f(g(\tau, z))$$

for $g \in G^J$, $f \in C^\infty(\mathbf{X})$. Then we have the

3.5.2 Remark. The differential operators introduced by the last remark obey the commutation rules

$$\pi_{m,k\pm2}(g) X_\pm^{m,k} = X_\pm^{m,k} \pi_{m,k}(g) \tag{3.23}$$

and

$$\pi_{m,k\pm1}(g) Y_\pm^{m,k} = Y_\pm^{m,k} \pi_{m,k}(g) \qquad \text{for all } g \in G^J. \tag{3.24}$$

Proof: The operators $\mathcal{L}_X$, $X \in \mathfrak{g}_{\mathbb{C}}^J$ are left invariant, i.e. we have

$$\mathcal{L}_X(\phi^{g'}) = (\mathcal{L}_X \phi)^{g'} \quad \text{with} \quad \phi^{g'}(g) := \phi(g'g).$$

For

$$\phi(g) = j_{k,m}(g) f(g(\mathbf{x}_0)) = j_{k,m}(g) f(\mathbf{x})$$

we get

$$\phi^{g'}(g) = j_{k,m}(g'g) f(g'g(\mathbf{x}_0)) = j_{k,m}(g) j_{k,m}(g', \mathbf{x}) f(g'(\mathbf{x})),$$

and thus by the last remark

$$(\mathcal{L}_{X_\pm} \phi^{g'})(g) = j_{k\pm2,m}(g) X_\pm^{k,m} j_{k,m}(g', \mathbf{x}) f(g'(\mathbf{x})),$$

as well as

$$\begin{aligned}
(\mathcal{L}_{X_\pm} \phi)^{g'}(g) &= j_{k\pm2,m}(g'g) X_\pm^{k,m} f(g'(x)) \\
&= j_{k\pm2,m}(g) j_{k\pm2,m}(g', \mathbf{x}) X_\pm^{k,m} f(g'(\mathbf{x})).
\end{aligned}$$

The relation for $Y_\pm^{m,k}$ comes out the same way. $\qquad\qquad\square$

Now, we denote by

$$\mathbb{D}_{m;k',k}(\mathbf{X})$$

the space of differential operators D on $\mathbf{X} = \mathbf{H} \times \mathbb{C}$ with

$$\pi_{m,k} \circ D = D \circ \pi_{m,k} \tag{3.25}$$

and ask for the structure of $\mathbb{D}_{m;k',k}(\mathbf{X})$, in particular for $\mathbb{D}(\mathbf{X}) := D_{0;0,0}(\mathbf{X})$. This will be deduced from Helgason's theory of differential operators on homogenous spaces, e.g. in [He1] or [He2], which obviously may be extended from $\mathbb{R}$-valued to $\mathbb{C}$-valued functions.

The starting point is the fact that our space $\mathbf{X} = G^J(\mathbb{R})/K^J(\mathbb{R})$ is a **reductive coset space**: As remarked in 1.4 we have

$$\mathfrak{g}^J = \mathfrak{k}^J + \mathfrak{m}^J \qquad \text{with} \quad (\mathrm{Ad}(k))\mathfrak{m}^J \subset \mathfrak{m}^J \quad \text{for all } k \in K^J.$$

Thus, we may apply his results and we have by Theorem 10 in [He1] or Theorem 2.8 and Corollary 2.9 in [He2] the following description of the algebra $\mathbb{D}(\mathbf{X})$ of invariant differential operators on $\mathbf{X}$: Let $I(\mathfrak{m}_{\mathbb{C}}^J)$ denote the polynomials P in the symmetric algebra $S(\mathfrak{m}_{\mathbb{C}}^J)$ of $\mathfrak{m}_{\mathbb{C}}^J$ which are invariant under $\mathrm{Ad}_G(K^J)$. Then, one has a linear bijection

$$I(\mathfrak{m}_{\mathbb{C}}^J) \xrightarrow{\sim} \mathbb{D}(\mathbf{X}),$$
$$Q \longmapsto D_{\lambda(Q)}.$$

Here, for a basis $X_1, \ldots, X_n$ of $I(\mathfrak{m}_{\mathbb{C}}^J)$ and $f \in \mathcal{C}^\infty(\mathbf{X})$, $D_{\lambda(Q)}$ is given by

$$D_{\lambda(Q)}f(\mathbf{x}) = \Big(Q(\lambda_{t_1}, \ldots, \lambda_{t_n})\phi(g\exp(t_1 X_1 + \cdots + t_n X_n)) \Big)_{t=0},$$

where ϕ is the K^J-invariant lift of f, i.e. $\phi = f \circ pr$, and λ indicates the symmetrization

$$\lambda(Y_1, \ldots, Y_p) = \frac{1}{p!} \sum_{\sigma \in \mathcal{S}_p} Y_{\sigma(1)} \circ \cdots \circ Y_{\sigma(p)}.$$

Helgason adds the warning that the map $Q \mapsto D_{\lambda(Q)}$ is not in general multiplicative (even when $\mathbb{D}(\mathbf{X})$ is commutative): We have

$$D_{\lambda(Q_1 Q_2)} = D_{\lambda(Q_1)} D_{\lambda(Q_2)} + D_{\lambda(Q)}$$

where $Q \in I(\mathfrak{m}_{\mathbb{C}}^J)$ has degree $\deg Q_1 + \deg Q_2$. We reproduce a result from [Be1].

3.5.3 Proposition. $\mathbb{D}(\mathbf{X})$ *is a non-commutative algebra generated by* $D_{\lambda(P_i)}$ *(i = 1, ..., 4) with*

$$
\begin{aligned}
\lambda(P_1) &= (1/2)(Y_+ Y_- + Y_- Y_+), \\
\lambda(P_2) &= (1/2)(X_+ X_- + X_- X_+), \\
\lambda(P_3) &= (1/6)(Y_+^2 X_- + Y_+ X_- Y_+ + X_- Y_+^2), \\
\lambda(P_4) &= (1/6)(Y_-^2 X_+ + Y_- X_+ Y_- + X_+ Y_-^2).
\end{aligned}
$$

Proof: We have $\mathfrak{m}_{\mathbb{C}}^J = \langle Y_\pm, X_\pm \rangle_{\mathbb{C}}$ and $Q \in I(\mathfrak{m}_{\mathbb{C}}^J)$ may be unterstood as a polynomial in $Y_\pm, X_\pm$ with $(\mathrm{ad}\, Z_0)Q = (\mathrm{ad}\, Z)Q = 0$. Z_0 being in the center of $\mathfrak{g}_{\mathbb{C}}^J$, the first equality gives no condition. To evaluate the second equality, we can deduce from the multiplication table in $\mathfrak{g}_{\mathbb{C}}^J$ that

$$\mathrm{ad}(Z)(X_+^j Y_-^j X_+^l X_-^m) = (j - k + 2l - 2m)(Y_+^j Y_-^k X_+^l X_-^m) \tag{3.26}$$

holds. This gives zero for the four "basic" combinations

$$j = k = 1; \qquad l = m = 1; \qquad j = 2, \qquad m = 1; \qquad k = 2, \qquad l = 1,$$

where only the non-zero numbers are mentioned. These four quadruples exactly lead to the terms given in the proposition, and from (3.26) it follows that all $\mathrm{Ad}(K^J)$-invariant elements of $U(\mathfrak{m}_{\mathbb{C}}^J)$ may be algebraically combined from these four terms. $\qquad\square$

As examples, the G^J–invariant differential operators on **X** corresponding to the first two terms are given by

$$\Delta_0^{0,0} = Y_+^{0,0} Y_-^{0,0} + Y_-^{0,0} Y_+^{0,0} = \frac{1}{2y}(\partial_p^2 - 2x\partial_p\partial_q + (x^2 + y^2)\partial_q^2)$$

and

$$\Delta_1^{0,0} = X_+^{0,0} X_-^{0,0} + X_-^{0,0} X_+^{0,0} = (2y^2)(\partial_x^2 + \partial_y^2).$$

3.5.4 Remark. As discussed in [Be5] and [Be6] for each $c > 0$

$$\Delta = \Delta_0 + c\Delta_1$$

is the Laplace-Beltrami operator belonging to a G^J-invariant metric on the space $\mathbf{X} = \mathbf{H} \times \mathbb{C}$ given by

$$ds^2 = y^{-2}(dx^2 + dy^2) + (cy)^{-1}((x^2 + y^2)\, dp^2 + 2x dp\, dq + dq^2)$$

It is not difficult to extend this to the determination of $\mathbb{D}_{m,k',k}(\mathbf{X})$. At first, we have a completely analogous situation for $k' = k$.

3.5.5 Corollary. $\mathbb{D}_{m;k}(\mathbf{X}) = \mathbb{D}_{m;k,k}(\mathbf{X})$ *is generated by the differential operators associated to* $P_1, \ldots, P_4$ *realized as polynomials in* $X_\pm^{m,k}$ *and* $Y_\pm^{m,k}$.

And another look at the relation (3.26) in the proof of the proposition above shows that for instance for $k' = k + 2$ we have the following statement.

3.5.6 Corollary. *Each element of* $\mathbb{D}_{m;k+2,k}(\mathbf{X})$ *is got by symmetrization of* $(Y_+^{(m,k)})^2$, $X_+^{(m,k)}$ *and elements of* $\mathbb{D}_{m;k}(\mathbf{X})$.

For later use we define the differential operators

$$\Delta^{k,m} := Y_+^{k,m} Y_-^{k,m} + Y_-^{k,m} Y_+^{k,m}. \tag{3.27}$$

3.6 Representations induced from $\hat{N}^J$ and Whittaker models

We start again by the Iwasawa decomposition, this time in the form

$$G^J(\mathbb{R}) = \hat{N}^J(\mathbb{R})A^J(\mathbb{R})K(\mathbb{R}), \qquad (\hat{N}^J := N^J Z)$$

take the character $\psi^{m,n,r}$ for $m, n, r \in \mathbb{Z}$, $m \neq 0$ and $\hat{n}(x, q, \kappa) \in \hat{N}^J(\mathbb{R})$ defined by

$$\psi^{m,n,r}(\hat{n}(x, q, \kappa)) = e(m\kappa + nx + rq), \qquad e(u) = \exp(2\pi i u), \qquad (3.28)$$

and study now the induced representation

$$\pi^{m,n,r} := \operatorname{ind}_{\hat{N}^J}^{G^J} \psi^{m,n,r}$$

acting by right translation on the space $\mathcal{W}^{m,n,r}$ of measurable $\mathbb{C}$–valued functions W living on $G^J(\mathbb{R})$ with the properties

i) $\qquad W(\hat{n}g) = \psi^{m,n,r}(\hat{n})W(g) \qquad$ for all $\hat{n} \in \hat{N}^J$, $g \in G^J$,

ii) $\qquad \|W\|^2 := \displaystyle\int_{\hat{N}^J \backslash G^J} |W(\tilde{t}(y)(r(\theta), (\hat{p}, 0, 0)))|^2 \, d\mu_{\hat{N}^J \backslash G^J} < \infty.$

Here we use coordinates

$$g = \hat{n}(x, \hat{q}, \hat{\zeta})\tilde{t}(y)(r(\theta), \hat{p}, 0, 0))$$

related to the "old" S–coordinates by

$$
\begin{aligned}
\hat{q} &= q + px \\
\hat{p} &= py^{1/2} \\
\hat{\zeta} &= \zeta e(p(px + q)).
\end{aligned}
$$

and the quasi-invariant measure on $\hat{N}^J \backslash G^J$ given by

$$d\mu_{\hat{N}^J \backslash G^J} = y^{-5/2} \, dy \, d\theta \, d\hat{p}.$$

Then in these coordinates, an element $W \in \mathcal{W}^{m,n,r}$, called **Whittaker function of type** (m, n, r), is of the form

$$W(g) = \psi^{m,n,r}(\hat{n})F(y, \theta, \hat{p})$$

with

$$\|W\|^2 = \int |F(y, \theta, \hat{p})|^2 y^{-5/2} \, dy \, d\hat{p} \, d\theta < \infty.$$

Now, we look for subspaces $\mathcal{W}^{m,n,r}(\pi)$ such that the restriction of right translation ρ to the subspace is equivalent to a given representation π of G^J with central character ψ^m. This will be called a **Whittaker model of type** (n,r) for π.

There are several ways to get at the Whittaker models. As we have developed the infinitesimal method and the method of differential operators this far, we will use it again here. It is convenient to separate moreover the K-variable θ and take for $k \in \mathbb{Z}$ and fixed m, n, r

$$W(g) = c_k(x, \hat{q}, \hat{\kappa}, , \theta)\varphi_k(y, \hat{p}) \quad \text{with} \quad c_k(x, \hat{q}, \hat{\kappa}, \theta) = \psi^{m,n,r}(\hat{n})e^{ik\theta}.$$

3.6.1 Remark. Applying the G^J–leftinvariant differential operators $\mathcal{L}_X$ we get by a small computation

$$
\begin{aligned}
\mathcal{L}_{Z_0}W &= 2\pi m W & \mathcal{L}_Z W &= kW \\
\mathcal{L}_{Y_\pm}W &= c_{k\pm1}\varphi_k^\pm & \mathcal{L}_{X_\pm}W &= c_{k\pm2}\varphi_{k\pm},
\end{aligned}
$$

where we have for $\varphi = \varphi_k$

$$
\begin{aligned}
\varphi^\pm &= (1/2)\varphi_{\hat{p}} \mp (2\pi m\hat{p}y^{1/2} + \pi r y^{1/2})\varphi \\
\varphi_\pm &= (1/2)\hat{p}\varphi_{\hat{p}} + y\varphi_y \mp (2\pi(m\hat{p}^2 + r\hat{p}y^{1/2} + ny) - k/2)\varphi.
\end{aligned}
$$

Guided by the form of the automorphic factor, we normalize the φ_k and put

$$\varphi_k = \varphi_k^{(0)}\psi_k, \qquad \varphi_k^\pm = \varphi_k^{(0)}\psi_k^\pm, \qquad \varphi_{k\pm} = \varphi_k^{(0)}\psi_\pm$$

with

$$\varphi_k^{(0)}(y, \hat{p}) = y^{k/2}e^{-2\pi(m\hat{p}^2 + r\hat{p}y^{1/2} + ny)}$$

resp. with $\mu = 2\pi m$ and $N = 4mn - r^2$

$$\varphi_k^{(0)}(y, \hat{p}) = y^{k/2}e^{-\mu(((\hat{p}+(r/2m))y^{1/2})^2 + (N/(4m^2))y)}.$$

I.e., now we have

$$W(g) = e(m(\kappa + pz) + n\tau + rz)y^{k/2}e^{ik\theta}\psi_k$$

or

$$= j_{k,m}(g, (i,0))e(n\tau + rz)\psi_k$$

with

$$
\begin{aligned}
\psi^+ &= (1/2)\psi_{\hat{p}} - 2\mu(\hat{p} + (r/2m))y^{1/2}\psi, \\
\psi^- &= (1/2)\psi_{\hat{p}},
\end{aligned}
$$

and

$$\psi_+ = (1/2)\hat{p}\psi_{\hat{p}} + (k - 2\mu(\hat{p}^2 + (r/(2m))\hat{p}y^{1/2} + (n/(2m))y)\psi,$$
$$\psi_- = (1/2)\hat{p}\psi_{\hat{p}} + y\psi_y.$$

From here, we immediately get a statement about the **existence and uniqueness of the Whittaker model for the discrete series representations** $\pi_{m,k}^+$ of $G^J(\mathbb{R})$.

3.6.2 Proposition. For $m > 0$, $N = 4mn - r^2 > 0$ and $k \geq 2$ there is exactly one subspace $\mathcal{W}^{m,n,r}(\pi_{m,k}^+)$ contained in $\mathcal{W}^{m,n,r}$ such that the right regular representation ρ restricted to this space is equivalent to $\pi_{m,k}^+$. We have the same statement for $\pi_{m,k}^-$ with $N = 4mn - r^2 < 0$ and $k \geq 2$.

Proof: By Proposition 3.1.7 $\pi = \pi_{m,k}^+$ has a cyclic vector W_0 of lowest weight characterized by

$$\hat{\pi}(Y_-)W_0 = \hat{\pi}(X_-)W_0 = 0, \ \hat{\pi}(Z)W_0 = kW_0, \ \hat{\pi}(Z_0)W_0 = \mu W_0. \tag{3.29}$$

As we have here $\hat{\pi}(X) = \mathcal{L}_X$ for $X \in \mathfrak{g}_{\mathbb{C}}^J$, it is clear by the last remark and the formulae in the sequel that for $\psi_k = 1$

$$W(g) = j_{k,m}(g,(i,0))e(n\tau + rz)$$

is such a vector, and it is unique up to a constant factor. We further have

$$\|W\|^2 = \int\limits_0^\infty \int\limits_{-\infty}^\infty y^k e^{-2\mu((\hat{p}+(r/2m)y^{1/2})^2 + (N/(4m)^2)y)} y^{-5/2} d\hat{p} \, dy$$

$$= (1/2)m^{-1/2} \int\limits_0^\infty e^{-(\pi N/(2m))y} y^{k-5/2} \, dy < \infty$$

for $N, m > 0$, $k \geq 2$. Again by Proposition 3.1.7, $\pi = \pi_{m,k}^-$ has a cyclic vector W_0 characterized by

$$\hat{\pi}(Y_-)W_0 = \hat{\pi}(D_+)W_0 = 0, \ \hat{\pi}(Z)W_0 = (1 - k)W_0, \ \hat{\pi}(Z_0) = 2\mu W_0. \tag{3.30}$$

The action of D_+ is given here by

$$(\mathcal{L}_{X_+} + (2\mu)^{-1}\mathcal{L}_{Y_+}^2)W = (W/\psi_{1-k})(\psi_{(1-k)+} + (2\mu)^{-1}\psi_{1-k}^{++})$$

where the formulae for ψ_+ and ψ^+ combine to

$$D_+\psi := \psi_+ + (2\mu)^{-1}\psi^{++}$$
$$= (1/(8\mu))\psi_{\hat{p}\hat{p}} - (\hat{p}/2 + ry^{1/2}/(2m))\psi_{\hat{p}} + y\psi_y + (ay + b)\psi$$

with

$$a = -\pi N/m, \qquad b = 1/2 - k.$$

$\hat{\pi}(Y_-)W = 0$ translates into $\psi^- = 0$, i.e. $\psi_{\hat{p}} = 0$. Thus, ψ is a function depending alone on y, and $\hat{\pi}(D_+)W = 0$ is equivalent to

$$y\psi_y + (ay + b)\psi = 0.$$

Hence, we have up to a constant factor

$$\psi_{1-k} = \psi = y^{-b}e^{-ay} = y^{k-1/2}e^{\pi Ny/m}$$

and

$$\begin{aligned}
W(g) &= j_{k,m}(g,(i,0))e(n\tau + rz)y^{k-1/2}e^{\pi Ny/m}\\
&= e(m(\kappa + p(px + q)) + nx + r(q + px))e^{i(1-k)\theta}y^{k/2-1/2}\\
&\qquad\qquad\qquad\qquad \cdot e^{-2\pi m(\hat{p}+y^{1/2}r/(2m))^2 + \pi Ny/(2m)}.
\end{aligned}$$

Here the norm $\|W\|$ is seen to be finite for $N < 0$ and $k \geq 2$. $\qquad\square$

The treatment of the **principal series** is a little bit more subtle. By Proposition 3.1.6 a cyclic vector W_0 for $\pi = \pi_{m,s,\nu}$ is characterized by

$$\hat{\pi}(Y_-)W_0 = 0, \quad \hat{\pi}(Z)W_0 = (1/2 + \nu)W_0, \quad \hat{\pi}(Z_0)W_0 = \mu W_0 \qquad (3.31)$$

and

$$\hat{\pi}(D_-D_+)W_0 = \lambda W_0 \qquad \text{with} \quad \lambda = (1/4)(s^2 - (\nu + 1)^2) \tag{3.32}$$

For $W(g) = e(m(\kappa + pz) + n\tau + rz)y^{k/2}e^{ik\theta}\psi_k$, we have here $k = 1/2 + \nu$, $\mu = 2\pi m$, and $\hat{\pi}(Y_-)W_0 = 0$ makes again that $\psi = \psi_k$ is a function of y alone. The relation (3.32), by the formula for D_+ given in the proof of the last proposition and a similar formula for D_-, comes down in this case to

$$y^2\psi_{yy} + (ay^2 + by)\psi_y - (b + \lambda)\psi = 0,$$

where λ is as above and we have again $a = -\pi N/m$ and $b = k - 1/2$. Here we substitute

$$\psi(y) = e^{\sigma(y)}\chi(-ay)$$

and get for

$$\sigma = -(1/2)(ay + b\log y)$$

the equation

$$\chi'' + (-1/4 - b/(2ay) + (-(b/2)(1 + b/2) - \lambda)/(ay)^2)\chi = 0$$

i.e., substituting b and λ for $k = 1/2 + \nu$

$$\chi'' + (-1/4 + \nu/(-2ay) + ((1/4)(1 - s^2)/(ay)^2)\chi = 0.$$

In Whittaker-Watson [WW] p. 337 we find the equation for the confluent hypergeometric function $W = W_{k_W,m_W}(z)$

$$\frac{d^2W}{dz^2} + \left(-\frac{1}{4} + \frac{k}{z} + \left(\frac{1}{4} - m^2\right)z^{-2}\right)W = 0. \tag{3.33}$$

Hence, our equation may be identified with this equation for

$$k_W = \nu/2 \quad \text{and} \quad m_W = s/2,$$

where we have replaced the letters k, m from Whittaker-Watson by k_W and m_W to distinguish them from the letters already used and fixed in our context. Now, in [WW] p. 337 we find as independent solutions of (3.33) for small $|z|$ and $2m_W \notin \mathbb{Z}$

$$W(z) = z^{1/2 \pm m_W} e^{-z/2} \{1 + z*\}$$

and on p. 343 for $|z|$ large and $|\arg z| \leq \pi - \alpha < \pi$

$$W(z) = W_{k_W,m_W}(z) \sim e^{-z/2} z^{k_W} \{1 + \Sigma z^{-n}(\dots)\}$$

resp.

$$W(z) = W_{-k_W,m_W}(-z).$$

Putting all this together, we have for the cyclic vector W_0 for $\pi_{m,s,\nu}$ the form

$$W_0(g) = e(m(\kappa + pz) + n\tau + rz)e^{ik\theta}y^{1/4}e^{\pi Ny/(2m)}W_{\nu/2,s/2}(\pi Ny/m),$$

or equivalently

$$\begin{aligned}
W_0(g) &= e(m(\kappa + p(px + q) + nx + r(q + px)e^{ik\theta}y^{1/4} \\
&\quad \cdot e^{-2\pi m(p+r(2m))^2 y}W_{\nu/2,s/2}(\pi Ny/m).
\end{aligned}$$

Using the asymptotic behaviour of $W_{k_W,m_W}(z)$ we have the finiteness of the norm exactly for one fundamental solution, and we have as a final statement the **existence and uniqueness of the Whittaker model for the principal series representation** $\pi_{m,s,\nu}$:

3.6.3 Proposition. *For m and $N > 0$ (and similarly for $m, N < 0$) there is exactly one subspace $\mathcal{W}^{m,n,r}(\pi_{m,s,\nu})$ contained in $\mathcal{W}^{m,n,r}$ such that the right regular representation ρ restricted to this space is equivalent to $\pi_{m,s,\nu}$.*

As it is another example of beautiful analysis, it seems worthwile to present a **second approach** to these **Whittaker models** via an integral transformation. The models of the representations $\pi_{m,k}^{\pm}$ and $\pi_{m,s,\nu}$ in 3.2 were constructed by the induction procedure, i.e., they are spanned by smooth functions ϕ living on $G^J(\mathbb{R})$ and having the transformation property

$$\phi(\hat{n}(x,q,\kappa)\tilde{t}(y)g) = e^m(\kappa)y^{s_0}\phi(g) \tag{3.34}$$

with

$$s_0 = \begin{cases} \frac{1}{2}\left(s + \frac{3}{2}\right) & \text{for } \pi_{m,s,\nu}, \\ k - \frac{3}{2} & \text{for } \pi_{m,k}^{\pm}. \end{cases}$$

As usual, we get from these functions — at least formally — functions fulfilling the functional equation of a Whittaker function of type (m, n, r) by the map $I^{n,r}$

$$\phi \longmapsto W = W_\phi^{n,r}$$

with

$$W_\phi^{n,r}(g) = \int_{N^J} \phi(\tilde{w}^{-1}n_0 g)e(-nx_0 - rq_0)\,dn_0, \quad n_0 = n(x_0, q_0).$$

Later on it will become clear, that to get finite expressions here the integration over x_0 has to be taken as a certain path integral then to be specified. The easily verified commutation rule

$$\tilde{w}^{-1}n(x_0, q_0)t(y, p) = (t(y^{-1}), (0, p), p^2x_0 - 2pq_0)\tilde{w}^{-1}n(\tilde{x}_0, \tilde{q}_0),$$

where

$$\tilde{x}_0 = x_0 y^{-1}, \qquad \tilde{q}_0 = (q_0 - px_0)y^{-1/2},$$

leads to

$$W_\phi(t(y, p)) = \int \phi(t(y^{-1}), 0, p, 0)\tilde{w}^{-1}n(\tilde{x}_0, \tilde{q}_0))e^m(p^2x_0 - 2pq_0)$$
$$\cdot e(-n_1x_0 - r_1q_0)\,dn_0.$$

Changing variables by

$$x_0 = \tilde{x}_0 y, q_0 = \tilde{q}_0 y^{1/2} + p\tilde{x}_0 y, \qquad \text{i.e.} \qquad dn_0 = dx_0\,dq_0 = y^{3/2}\,d\tilde{x}_0\,d\tilde{q}_0$$

and using the transformation property (3.34), we get

$$W_\phi(t(y, p)) = y^{3/2-s_0} \iint \phi(\tilde{w}^{-1}n(x_0, q_0))e(-Ax_0 - Bq_0)dx_0\,dq_0$$

with

$$A = (mp^2 + rp + n)y = m(p + r/2m)^2y + Ny/(4m), \quad N = 4mn - r^2,$$
$$B = 2m(p + r/2m)y^{1/2},$$

or

$$W_\phi(t(y, p)) = y^{3/2-s_0}\widehat{\phi^w}(AB)$$

where $\widehat{\phi^w}$ denotes the Fourier transform of

$$\phi^w(x, q) := \phi(\tilde{w}^{-1}n(x, q)).$$

Using again the fact that the representations π considered here are cyclic and generated by functions realizing $\nu_0 \otimes w_0$ resp. $\nu_0 \otimes w_{1/2+\nu}$, the existence of Whittaker models may be proved by showing that the map $\phi \mapsto W_\phi$ makes sense for certain functions $\phi = \phi_1$ with the property (3.34) and defined by

$$\phi_1(g) = e^m(\kappa)e^{il\theta}y^{so}e^m(pz)$$

and has an image $W_{\phi_1} \neq 0$. To calculate W_{ϕ_1}, we start by the nasty commutation rule

$$\tilde{w}^{-1}n(x_0, q_0) = (n(x_1)t(y_1)r(\theta_1), \ (p_1, q_1, \kappa_1))$$

with

$$x_1 + iy_1 = \frac{-1}{i + x_0} = \frac{i - x_0}{1 + x_0^2}, \qquad e^{i\theta_1} = \frac{x_0 - i}{(1 + x_0^2)^{1/2}} = \left(\frac{x_0 - i}{x_0 + i}\right)^{1/2},$$

$$p_1 = -q_0, q_1 = 0, \ \kappa_1 = 0$$

Hence we have

$$\phi_1^w(x_0, q_0) = (x_0^2 + 1)^{-(s_0 + l/2)}(x_0 - i)^l e^m(-q_0^2/(x_0 + i))$$

and therefore

$$W_{\phi_1}(t(y, p)) = y^{3/2 - s_0} \iint (x_0^2 + 1)^{-(s_0 + l/2)}(x_0 - i)^l$$
$$\cdot e(-mq_0^2/(x_0 + i) - Ax_0 - Bq_0) \, dx_0 \, dq_0.$$

The usual Fourier transformation formula

$$\int_{\mathbb{R}} e^{-\pi(a+ib)x^2}e^{-2\pi ixy} \, dx = (a + ib)^{-1/2}e^{-\pi y^2/(a+ib)}, \qquad a, b \in \mathbb{R}, \ a > 0,$$

may be applied here with

$$x = q_0, \qquad a + ib = 2m(1 + ix_0)/(1 + x_0^2)$$

to give

$$W_{\phi_1}(t(y, p)) = y^{3/2 - s_0}(2mi)^{-1/2}e^{-\pi B^2/(2m)}$$
$$\cdot \int (x_0^2 + 1)^{-(s_0 + l/2 - 1/2)}(x_0 - i)^{l - 1/2} \cdot e^{-\pi iNx_0y/(2m)} \, dx_0 \tag{3.35}$$

Integrals such as the one appearing in this expression describe classical Whittaker functions and show up in the literature in different places. For instance in [Wa1] pp. 23–24 Waldspurger discusses on his way to Whittaker models for the metaplectic group a function

$$W_n(\alpha, \eta, \theta) = \alpha^{1-s} e^{in\theta} \int_{\Gamma_c} (1+b^2)^{-(s+1+n)/2} (i-b)^n e^{-2\pi ic(b\alpha^2 - \eta\alpha^2)} \, dp$$

where

$$\alpha, \eta, \theta, c \in \mathbb{R}, \ \alpha > 0, c \neq 0, s \in \mathbb{C}, n \in 1/2 + \mathbb{Z}$$

and Γ_c is a path like in the sketch.

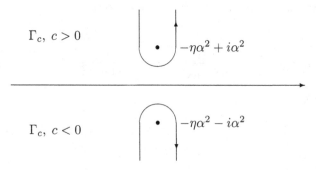

$\Gamma_c, \ c > 0$ $-\eta\alpha^2 + i\alpha^2$

$\Gamma_c, \ c < 0$ $-\eta\alpha^2 - i\alpha^2$

i) For $s = -1/2 + 2\nu$ and $n = s + 1 = 1/2 + 2\nu$, $\nu \in \mathbb{Z}$, he gives for

$$W_n(1,0,0) = i^n \int_{\Gamma_c} (1 - ib)^{-(s+1)} e^{-2\pi icb} \, db \qquad\qquad (3.36)$$

the expression

$$W_n(1,0,0) = |c|^{1/2} (2\pi c)^s e^{-2\pi c} \Gamma(-s) 2 \sin(\pi(s+1)) i^n. \qquad\qquad (3.37)$$

Here the factor $|c|^{1/2}$ seems to us to be superfluous, but in any case the function is not zero for a half integer s.

ii) For $s \notin 1/2\mathbb{Z}$ and $n = 1/2$ Waldspurger remarks in [Wa1] on p. 94 that $W_n(\alpha, 0, 0)$ is not identically zero.

Now, these results may be used for the discussion of the map $\phi \mapsto W_\phi$ resp. the calculation of W_{ϕ_1} in the following way. At first we are led to fix the integration over x_0 in the integral transform W_ϕ as a path integral over Γ_c as in the sketch above. Then, we realize case by case a lowest weight vector of the different types of representations π.

Case $\pi_{k,m}^+$

By Corollary 3.3.2 we have to take ϕ_1 with $s_0 = k/2$ and $l = k$. Thus, the integral in (3.35) specializes to

$$\int_{\Gamma_c} (x_0 + i)^{-k+1/2} e^{-2\pi i c x_0} \, dx_0, \quad c = Ny/(4m), \quad \text{for} \quad s = k - 3/2 \tag{3.38}$$

i.e. essentially the integral for W_n in (3.36) for $s = k - 3/2$. This already shows that (3.35) gives a nontrivial representation of a lowest weight vector by a Jacobi-Whittaker function. And introducing (3.38) evaluated by (3.37) (without $|c|^{1/2}$), we get up to trivial nonzero factors exactly the function

$$W(t(y,p)) = y^{k/2} e^{-2\pi m((p+r/2m)^2 y)} e^{-\pi Ny/(2m)},$$

which is the specialization of the function $W(g)$ appearing in the proof of Proposition 3.6.3.

Case $\pi_{m,k}^-$

Here we have to take

$$\phi_1(g) = e^m(\kappa) e^m(pz) y^{k/2} e^{i(1-k)\theta},$$

i.e. $s_0 = k/2$ and $l = 1 - k$. Then (3.35) specializes to

$$W_{\phi_1}(t(y,p)) = y^{3/2-k/2} (2im)^{-1/2} e^{-2\pi m(p+r/(2m))^2 y}$$

$$\cdot \int_{\Gamma_c} (x_0 - i)^{1/2-k} e^{-2\pi i (Ny/(4m)) x_0} \, dx_0.$$

For the integral we can again take over Waldspurger's result: Comparison with (3.36) shows that this time we have to put $b = -x_0$, $c = -Ny/(4m)$ and $s = k - 3/2$. Using (3.37) we then get up to a trivial nonzero factor the function

$$W(t(y,p)) = y^{k/2} e^{-2\pi m(p+r/(2m))^2 y} e^{\pi Ny/(2m)}$$

which is again a special value of the function $W(g)$ appearing in the proof of Proposition 3.6.3 for $\pi_{m,k}^-$.

Case $\pi_{m,s,\nu}$

We start by the cyclic spherical (or nearly spherical) vector

$$\phi_1(g) = e^m(k + pz) y^{(s+3/2)/2} e^{il\theta}, \qquad l = \nu + 1/2, \ \nu = \pm 1/2,$$

i.e., we have here

$$s_0 = s/2 + 3/4, \qquad l = \nu + 1/2,$$

and (3.35) specializes to

$$W_{\phi_1}(t(y,p)) = y^{3/4-s/2}(2mi)^{-1/2}e^{-2\pi m(p+r/2m)^2 y}$$

$$\cdot \int |x_0^2 + 1|^{-s-1+\nu}(x_0 + i)^{-\nu}e^{-2\pi i N y x_0/(4m)}\, dx_0.$$

$$(3.39)$$

The integral in this expression may be identified with the integral given by Waldspurger und thus, by his result cited above in ii), we have a nontrivial Whittaker function again. But this still may be pursued a bit further. In [Ja] p. 283 Jacquet introduces the function

$$W_{k_J}(u, s_J) = u^{s_J} \int_{-\infty}^{\infty} |t + iu|^{2k_J - 2s_J}(t + iu)^{-2k_J}e^{-it}\, dt$$

with

$$u = \pi N/(2m), \qquad k_J = \nu/2 \text{ and } s_J = s/2 + 1/2.$$

Introducing this function into our expression, we have up to constant nonzero factors

$$W_{\phi_1}(t(y,p)) \sim y^{1/4}W_{\nu/2}(\pi N/2m, s/2+1/2) \text{ for } \operatorname{Re} s \gg 0$$

and, using moreover [Ja] (4.2.17), we come back to the classical Whittaker function $W_{k,m}$ and get

$$W_{\phi_1}(t(y,p)) \sim y^{1/4}e^{-2\pi m(p+r/2m)^2 y}W_{\nu/2,-s/2}(\pi N y/m),$$

hence again a special value of the function from the proof of Proposition 3.6.3.

The following statement summarizes the content of this section. It is of some importance for the definition of an L-factor "at infinity".

3.6.4 Corollary. *Let* $n, r \in \mathbb{Z}$ *and* $N := 4mn - r^2$. *We have existence and uniqueness of the Whittaker models* $\mathcal{W}^{n,r}(\pi)$ *for*

$$\pi = \pi_{m,k}^+ \qquad \text{with} \quad mN > 0 \qquad (k \geq 1),$$

$$\pi = \pi_{m,k}^- \qquad \text{with} \quad mN < 0 \qquad (k \geq 1),$$

$$\pi = \pi_{m,s,\nu} \qquad \text{with} \quad mN > 0.$$

In all three cases we have a distinguished cyclic element $W_0 \in \mathcal{W}^{n,r}(\pi)$, namely

$$W_0(g) = e(m(\kappa + pz) + n\tau + rz)e^{ik\theta}y^{k/2} \qquad \qquad \text{for } \pi_{m,k}^+,$$

$$W_0(g) = e(m(\kappa + pz) + n\tau + rz)e^{i(1-k)\theta}y^{k/2}e^{\pi N y/m}$$

$$= e(m(\kappa + p(px+q)) + nx + r(q + px))e^{i(1-k)\theta}y^{k/2}$$

$$\cdot e^{-2\pi m(p+r/(2m))^2 y + \pi N y/(2m)} \qquad \text{for } \pi_{m,k}^-,$$

and with the classical Whittaker function $W_{m,k}(z)$ from [WW], p. 337,

$$W_0(g) = e(m(k + p(px + q)) + nx + r(q + px))e^{i(\nu+1/2)\theta}y^{1/4}$$

$$\cdot e^{-2\pi m(p+r/(2m))^2 y}W_{\nu/2,s/2}(\pi Ny/m) \qquad \text{for } \pi_{m,s,\nu}.$$

In all three cases W_0 is the image

$$W_0 = W_{\phi_0}^{n,r} =: I^{n,r}\phi_0$$

of the corresponding cyclic vector $\phi_0 \in \mathcal{B}_\pi$ from the model coming from the induction procedure, i.e., ϕ_0 and W_0 are lowest resp. dominant weight vectors for $\pi_{m,k}^\pm$ and spherical resp. nearly spherical vectors for $\pi_{m,s,\nu}$.

For the sake of completeness, we mention that there is still another way leading to the Whittaker models, closely related to the last discussion and of great importance in the non archimedean case. Namely, for a representation π with representation space V_π consisting of smooth functions ϕ we look at a **Whittaker functional** $l^{n,r}$, defined as a continuous linear map

$$l^{n,r} : V_\pi \longrightarrow \mathbb{C}$$

with the property

$$l^{n,r}(\pi(n)\phi) = \psi^{n,r}(n)l^{n,r}(\phi). \tag{3.40}$$

Then the associated Whittaker model is given by right translation upon the space

$$\mathcal{W}^{n,r} = \{g \mapsto l^{n,r}(\pi(g)\phi) : \phi \in V_\pi\}.$$

In our case the Whittaker functional is uniquely given by

$$l^{n,r}(\phi) = \int_{N^J} \phi(\tilde{w}^{-1}n)\overline{\psi^{n,r}(n)}\,dn$$

where the integration has to be taken carefully as explained above.

4

The Space $L^2(\Gamma^J \backslash G^J(\mathbb{R}))$ and its Decomposition

In the last chapter the induction procedure presented in 2.1 was exploited, starting by the subgroups B^J, K^J and $\tilde{N}^J$. Now, another, albeit rather trivial, way to use this again is to take the discrete subgroup

$$\Gamma'^J = \mathrm{SL}_2(\mathbb{Z}) \ltimes \mathbb{Z}^2$$

of $G'^J(\mathbb{R})$ or equivalently the subgroup $\Gamma^J = \mathrm{SL}_2(\mathbb{Z}) \ltimes H(\mathbb{Z})$ of $G^J(\mathbb{R})$, and in each case the trivial representation id, and induce from here, i.e., to study the representation $\mathrm{ind}_{\Gamma^J}^{G'^J(\mathbb{R})}(\mathrm{id})$ given (in the "second realization") by right translation ρ on the space

$$\mathcal{H} = L^2(\Gamma'^J \backslash G'^J(\mathbb{R})).$$

We will collect in this chapter some material (prepared in [BeBö] and [Be3]) about the decomposition of this representation into a cuspidal and a continuous part. We hope to give an impression of the theory even if we restrict to some main points, for instance leaving aside the possibility to replace here Γ'^J by some, say, congruence subgroup

$$\Gamma'^J(N, N') := \Gamma(N) \ltimes (N'\mathbb{Z})^2.$$

Not striving for the same completeness as in the other chapters, we will at least introduce the standard objects showing up here, i.e., the Jacobi forms, Jacobi Eisenstein series and more general automorphic forms. In this chapter G'^J will stand for $G'^J(\mathbb{R})$ and G^J for $G^J(\mathbb{R})$.

4.1   Jacobi forms and more general automorphic forms

As $K^J = SO(2) \times S^1$ is a commutative compact subgroup of $G'^J(\mathbb{R})$, we have the decomposition of ρ related to the characters $\chi_{m,k}$ of K^J, namely

$$\mathcal{H} = L^2(\Gamma'^J\backslash G'^J) = \bigoplus_{m\in\mathbb{Z}} \mathcal{H}_m$$

with

$$\mathcal{H}_m = \{\phi \in \mathcal{H}: \quad \phi(g\zeta) = \zeta^m\phi(g) \quad \text{for all} \quad \zeta \in S^1, \ g \in G^J\}$$

and

$$\mathcal{H}_m = \bigoplus_{k\in\mathbb{Z}} \mathcal{H}_{m,k}$$

with

$$\mathcal{H}_{m,k} = \{\phi \in \mathcal{H}_m: \quad \phi(gr(\theta)) = e^{ik\theta}\phi(g) \quad \text{for all} \quad r(\theta) \in SO(2), \ g \in G^J\}.$$

By the discussion in 3.4, we have the notion of the "canonical automorphic factor" $j_{k,m}$ and with it the distinction of elements $\phi = \phi_f \in \mathcal{H}_{m,k}$, which may be interpreted as lifts of certain holomorphic functions f living on

$$\mathbf{X} = \mathbf{H} \times \mathbb{C} = G'^J/K^J.$$

We take this as a motivation to repeat here the usual definition of the Jacobi forms from [EZ] and to discuss moreover some generalizations (even if part of these won't appear in the decomposition of $\mathcal{H}$).

Holomorphic Jacobi forms

The canonical automorphic factor for the Jacobi group, as described in 3.4, goes back to Satake and has in the EZ-coordinates

$$g = n(x)t(y)r(\theta)(\lambda, \mu, \zeta) \in G'^J$$

the form

$$j_{k,m}(g, (\tau, z)) = \zeta^m e^m\left(-\frac{c(z + \lambda\tau + \mu)^2}{c\tau + d} + \lambda^2\tau + 2\lambda z + \lambda\mu\right)(c\tau + d)^{-k}.$$

We take over the definition from [EZ] p. 9:

4.1.1 Definition. *A* **Jacobi form of weight** k *and* **index** m $(k, m \in \mathbb{N})$ *is a complex valued function* f *on* $\mathbf{H} \times \mathbb{C}$ *satisfying*

   i) $(f|_{k,m}[\gamma])(\tau, z) := j_{k,m}(\gamma, (\tau, z))f(\gamma(\tau, z)) = f(\tau, z)$ *for all* $\gamma \in \Gamma'^J$

   ii) *f is holomorphic*

iii) f has a Fourier development of the form

$$f(\tau, z) = \sum_{\substack{n, r \in \mathbb{Z} \\ 4mn - r^2 \geq 0}} c(n, r) e(n\tau + rz).$$

*f is called a **cusp form**, if it satisfies moreover*

iii') $c(n, r) = 0$ unless $4mn > r^2$.

The vector spaces of all such functions f are denoted by $J_{k,m}$ resp. $J_{k,m}^{\text{cusp}}$. They are finite dimensional by Theorem 1.1 of [EZ]. One could also define Jacobi forms for subgroups of Γ'^J of finite index, but we do not need this here.

As an easy consequence of the transformation law i) one has for the Fourier coefficients the following fundamental result ([EZ] Theorem 2.2)

$$c(n, r) \text{ depends only on } N = 4mn - r^2 \text{ and on } r \bmod 2m \qquad (4.1)$$

As already mentioned in the introduction, there is a lot of work done using these Jacobi forms. We will here not go into this but only indicate the **characterization of Jacobi forms as functions on** $G'^J(\mathbb{R})$ (as in [BeBö] 5.).

4.1.2 Proposition. $J_{k,m}$ *is isomorphic to the space $\mathcal{A}_{m,k}$ of complex functions $\phi \in \mathcal{C}^\infty(G'^J)$ with*

i) $\phi(\gamma g) = \phi(g)$ for all $\gamma \in \Gamma'^J$

ii) $\phi(g r(\theta, \zeta)) = \phi(g) \zeta^m e^{ik\theta}$ for all $r(\theta, \zeta) \in K^J$

iii) $\mathcal{L}_{Y_-} \phi = \mathcal{L}_{X_-} \phi = 0$

iv) for all $M \in \mathrm{SL}_2(\mathbb{Z})$ the function

$$g \longmapsto \phi(g) y^{-k/2}$$

is bounded in domains of type $y > y_0$.

$J_{k,m}^{\text{cusp}}$ *is isomorphic to the subspace $\mathcal{A}_{m,k}^0$ of $\mathcal{A}_{m,k}$ with*

iv') the function $g \mapsto \phi(g)$ is bounded.

Proof: As in 3.4, for each $k, m \in \mathbb{N}_0$, the automorphic factor $j_{k,m}$ defines a lifting $\varphi_{k,m}$ from functions f living on $\mathbf{H} \times \mathbb{C}$ to functions ϕ living on G'^J by

$$f \xrightarrow{\varphi_{k,m}} \phi_f \quad \text{with } \phi_f(g) = f(g(i, 0)) j_{k,m}(g, (i, 0))$$

$$= f(x, y, p, q) \zeta^m e^m(pz) e^{ik\theta} y^{k/2},$$

where g is meant in the S–coordinates $(x, y, \theta, p, q, \zeta)$ and the letter f denotes the function in the four real variables x, y, p, q, which, when holomorphic as a function of

$$\tau = x + iy \quad \text{and} \quad z = p\tau + q,$$

is also denoted by $f(\tau, z)$. Now, $\varphi_{k,m}$ identifies the space $\mathcal{F}^\Gamma_{k,m}$ of functions f on $\mathbf{H} \times \mathbb{C}$ satisfying the transformation formula i) of the Jacobi forms with the set $\mathcal{F}^\Gamma_{k,m}(G'^J)$ of functions $\phi : G'^J \to \mathbb{C}$ satisfying i) and ii). The equivalence of the holomorphy of f and the equations

$$\mathcal{L}_{Y_-} \phi = \mathcal{L}_{X_-} \phi = 0$$

for $\phi = \phi_f$ are immediate from the formulae for $\mathcal{L}_{X_-}$ and $\mathcal{L}_{Y_-}$, as already remarked in 3.5. The condition iii) resp. iii') in the definition of the Jacobi forms and the condition iv) resp. iv') in the proposition are equivalent by the following standard fact.

4.1.3 Lemma. *For a holomorphic function $f : \mathbf{H} \times \mathbb{C} \to \mathbb{C}$ with Fourier expansion*

$$f(\tau, z) = \sum_{n,r \in \mathbb{Z}} c(n,r) e(n\tau + \tau z)$$

the condition

    a) *$c(n,r) = 0$ for all n, r with $N = 4mn - r^2 < 0$*

is equivalent to

    b) *For all positive real numbers y_0 and p_0 the function*

$$f(\tau, z) e^m(pz)$$

    is bounded in domains of type

$$\{(\tau, z) \in \mathbf{H} \times \mathbb{C} : y \geq y_0, \, |p| \leq p_0\}$$

    (where as before $\tau = x + iy$ and $z = p\tau + q$).

Proof: For all n, r and $y > 0$, $\eta \in \mathbb{R}$ we have with $\tau = x + iy$, $z = p\tau + q = \xi + i\eta$

$$c(n,r) e^{-2\pi(ny+r\eta)} = \iint\limits_{(\mathbb{R}/\mathbb{Z})^2} f(\tau, z) e^{-2\pi i(nx + r\xi)} \, d\xi \, dx. \tag{4.2}$$

If $N = 4mn - r^2 < 0$, then there is a $p_1 \in \mathbb{R}$ with

$$n + rp_1 + mp_1^2 = \lambda < 0.$$

(4.2) for $\eta = p_1 y$ gives after multiplication by $e^{-2\pi m y p_1^2}$

$$c(n,r)e^{-2\pi y\lambda} = \iint\limits_{(\mathbb{R}/\mathbb{Z})^2} f(\tau, z)e^{-2\pi i(nx+r\xi)-2\pi m y p_1^2}\, d\xi\, dx.$$

and hence

$$|c(n,r)|e^{-2\pi y\lambda} \le \iint\limits_{(\mathbb{R}/\mathbb{Z})^2} |f(x+iy, \xi+ip_1 y)|e^{-2\pi m y p_1^2}\, dx\, d\xi$$

$$= \iint\limits_{(\mathbb{R}/\mathbb{Z})^2} \Big| f(x+iy, \xi+p_1(x+iy))e^m(p_1(\xi+p_1(x+iy))) \Big|\, dx\, d\xi.$$

The boundedness condition b) now implies

$$|c(n,r)|e^{-2\pi y\lambda} \le L$$

for all $y \ge y_0$, where $L > 0$ depends only on p_1 and some $y_0 > 0$. Since $\lambda < 0$, this implies $c(n,r) = 0$.

Assume conversely a) to be fulfilled. It is a well known fact that the series

$$\sum_{n,r} |c(n,r)e(n\tau + rz)|$$

converges uniformly on compact sets, and hence the definition

$$A(y,p) = \sum_{n,r} |c(n,r)|e^{-2\pi y(mp^2+rp+n)}$$

$$= e^{-2\pi y m p^2} \sum_{n,r} |c(n,r)e(niy + rpiy)|$$

makes sense and defines a real-valued continuous function on $\mathbb{R}_{>0} \times \mathbb{R}$. Now for $m > 0$ and $4mn - r^2 \ge 0$

$$mp^2 + rp + n \ge 0 \qquad \text{for all } p \in \mathbb{R},$$

and thus

$$A(y,p) \le A(y_0, p) \qquad \text{for all } y \ge y_0.$$

By continuity there is a constant $L > 0$ such that

$$A(y_0, p) \le L \qquad \text{for all } p \in \mathbb{R} \text{ with } |p| \le p_0.$$

On the set

$$\{(\tau, z) \in \mathbf{H} \times \mathbb{C} : y \ge y_0, |p| \le p_0\}$$

we thus have the chain of inequalities

$$|f(\tau, z)e^m(pz)| \le A(y,p) \le A(y_0, p) \le L. \qquad \square$$

The conditions ii) and iii) in the last proposition show that each $\phi \in \mathcal{A}_{m,k}$ qualifies as a candidate for a lowest weight vector of a discrete series representation $\pi^+_{m,k}$. Before going deeper into this, let us to some extend follow the observation that this should nourish the expectation to have a similar picture for the other types of representations $\pi^-_{m,k}$ and $\pi_{m,s,\nu}$.

Skoruppa's skew-holomorphic Jacobi forms

While studying certain general theta functions, Skoruppa introduced in [Sk2] p. 179 parallel to the definition of the space $J_{k,m}$ a space $J^*_{k,m}$ of **skew-holomorphic Jacobi forms** f of weight k and index m $(k, m \in \mathbb{N})$ as the space of smooth functions $f : \mathbf{H} \times \mathbb{C} \to \mathbb{C}$ satisfying

i) $f|^*_{k,m}[\gamma] = f$ for all $\gamma \in \Gamma'^J$

ii) $\partial_{\bar{z}} f = (8\pi i m \partial_\tau + \partial_z^2) f = 0$

iii) f has a Fourier development of the form

$$f(\tau, z) = \sum_{\substack{n,r \in \mathbb{Z} \\ 4mn - r^2 \leq 0}} c(n,r) e(n\tau + iy(r^2 - 4mn)/(2m) + rz).$$

$$(4.3)$$

Here again one has

$$
\begin{aligned}
\tau &= x + iy,\ z = p\tau + q = \xi + i\eta \\
\partial_\tau &= (1/2)(\partial_x - i\partial_y),\ \partial_z = (1/2)(\partial_\xi - i\partial_\eta),\ \partial_{\bar{z}} = (1/2)(\partial_\xi + i\partial_\eta)
\end{aligned}
$$

and the slash operator $|^*_{k,m}$ is given, slightly misusing the notation $f(\tau, z) = f(x, y, \xi, \eta)$, by

$$\left(f\Big|^*_{k,m}[g]\right)(\tau, z) = f(g(\tau, z)) j^*_{k,m}(g, (\tau, z))$$

with the automorphic factor

$$j^*_{k,m}(g, (\tau, z)) = j_{0,m}(g, (\tau, z))(c\bar{\tau} + d)^{-k+1}|c\tau + d|^{-1}.$$

As above in Proposition 4.1.2 we can lift these functions f to the group, this time by the lifting

$$
\begin{aligned}
f \longmapsto \phi_f \quad \text{with} \quad \phi_f(g) &= f(g(i, 0)) j^*_{k,m}(g, (i, 0)) \\
&= f(\tau, z) \zeta^m e^m(pz) e^{i(1-k)\theta} y^{k/2}.
\end{aligned}
$$

and by a slightly more difficult but similar proof (see [Be4]) we get

4.1.4 Proposition. $J^*_{k,m}$ is isomorphic to the space $\mathcal{A}^*_{m,k}$ of complex functions $\phi \in \mathcal{C}^\infty(G'^J)$ with

i) $\phi(\gamma g) = \phi(g)$ for all $\gamma \in \Gamma'^J$

ii) $\phi(gr(\theta, \zeta)) = \phi(g)\zeta^m e^{i(1-k)\theta}$ for all $r(\theta, \zeta) \in K^J$

iii) $\mathcal{L}_{Y_-}\phi = (4\pi m \mathcal{L}_{X_+} + \mathcal{L}_{Y_+}^2)\phi = 0$

iv) $\phi(g)y^{-k/2}$ is bounded in domains of type $y > y_0$.

Comparing this with Proposition 3.1.7, we see that each $\phi \in \mathcal{A}_{m,k}^*$ may be thought of as a cyclic vector for the representation $\pi_{m,k}^-$.

4.1.5 Remark. In [Sk2] the Fourier development is given a form which looks symmetrical for holomorphic and skew-holomorphic Jacobi forms. Let $f \in J_{k,m}$ have a Fourier development like in Definition 4.1.1 iii). By (4.1) the definition

$$C(\Delta, r) := c\left(\frac{r^2 - \Delta}{4m}, r\right) \qquad \text{for } \Delta \in -\mathbb{N}_0 \text{ and } r^2 \equiv \Delta \bmod 2m$$
$$(4.4)$$

makes sense. f can therefore be written as

$$
\begin{aligned}
f(\tau, z) &= \sum_{\substack{\Delta \leq 0}} \sum_{\substack{n, r \in \mathbb{Z} \\ \Delta = r^2 - 4mn}} c(n, r)e(n\tau + rz) \\
&= \sum_{\substack{\Delta \leq 0}} \sum_{\substack{r \in \mathbb{Z} \\ \Delta \equiv r^2 \bmod 4m}} C(\Delta, r)e\left(\frac{r^2 - \Delta}{4m}\tau + rz\right).
\end{aligned}
\qquad (4.5)
$$

Now, if $f^* \in J_{k,m}^*$ is a skew-holomorphic form with Fourier development (4.3), we make a similar definition as (4.4), with only $\Delta \in -\mathbb{N}_0$ replaced by $\Delta \in \mathbb{N}_0$, and arrive at

$$
f^*(\tau, z) = \sum_{\substack{\Delta \leq 0}} \sum_{\substack{r \in \mathbb{Z} \\ \Delta \equiv r^2 \bmod 4m}} C(\Delta, r)e\left(\frac{r^2 - \Delta}{4m}x + i\frac{r^2 + |\Delta|}{4m}y + rz\right).
\qquad (4.6)
$$

Of course, one could write Δ instead of $|\Delta|$ in this formula. The reason for prefering the absolute value is that if we change the summation from $\Delta \geq 0$ to $\Delta \leq 0$, then we get exactly the Fourier development (4.5) of f.

Maaß-Jacobi forms

It is tempting to try to generalize the Maaß wave forms for our case and moreover discuss the general notion of an automorphic form as for instance in Borel's article [B]. All this should be done in such a way that Arakawa's Eisenstein series in [Ar] and their generalizations in [Be3] fit in, and cyclic vectors for $\pi_{m,s,\nu}$ appear. We resume here and generalize a bit a discussion from [Be1] and [BeBö].

As we have developed in 3.5 the notion of a G'^J-invariant Laplace operator $\Delta_{\mathbf{H} \times \mathbb{C}}$ on $\mathbf{H} \times \mathbb{C}$ it is easy to propose the following

4.1.6 Definition. *A smooth function $f : \mathbf{H} \times \mathbb{C} \to \mathbb{C}$ is called a* **Maaß-Jacobi**
form if

   i) f is Γ^J-invariant

   ii) f is a $\Delta_{\mathbf{H} \times \mathbb{C}}$ eigenfunction

   iii) f fulfills a boundedness condition, say, f is of polynomial growth.

A variant of this would be using the operators $X_{\pm}^{m,k}$, $Y_{\pm}^{m,k}$ commuting with
$|_{k,m}$ and introduced in 3.5

4.1.7 Definition. *$f \in \mathcal{C}^\infty(\mathbf{H} \times \mathbb{C})$ is called a (k,m)-***Maaß-Jacobi form*** if*

   i) $f|_{k,m}[\gamma] = f$ for all $\gamma \in \Gamma^J$.

   ii) $\Delta^{k,m} f = \lambda f$ for some $\lambda \in \mathbb{C}$.

   iii) $f y^{-k/2}$ is bounded in domains of type $y > y_0$.

Arakawa's Eisenstein series in [Ar] and their generalizations in [Be3], to be discussed later on, are examples for these forms with $\lambda \neq 0$, and the holomorphic
Jacobi forms with $\lambda = 0$.

We may try to compare this with the **general notion of an automorphic form** as
defined in Borel [B] for a reductive group. Here we have the problem that the
second condition in our definitions, the analycity condition, usually is defined
using the center $\mathfrak{z}(\mathfrak{g}_\mathbb{C})$ of the universal enveloping algebra $U(\mathfrak{g}_\mathbb{C})$. And in our
case the center of $U(\mathfrak{g}_\mathbb{C}^J)$ is too small to single out anything useful. Now, the
definitions above propose to take the Laplacian of a G^J left invariant metric as
the operator to single out eigenfunctions ϕ on G^J. But we have a better more
invariant choice:

For fixed m, we defined in 3.1 the localization $U(\mathfrak{g}_\mathbb{C}^J)'$ of $U(\mathfrak{g}_\mathbb{C}^J)$ and found a
Casimir operator

$$C := D_+ D_- + D_- D_+ + (1/2)\Delta_1^2$$

which operates by multiplication with

$$\lambda = \begin{cases} \dfrac{1}{2}(s^2 - 1) & \text{for } \hat{\pi}_{m,s,\nu}, \\[2mm] \dfrac{1}{2}\left(k - \dfrac{1}{2}\right)\left(k - \dfrac{5}{2}\right) & \text{for } \hat{\pi}_{m,k}^{\pm}. \end{cases}$$

Realizing this C as a differential operator $\mathcal{L}_C$ as in 1.4, we now propose

4.1.8 Definition. *An automorphic form of type m, k, λ^* is a complex function
$\phi \in \mathcal{C}^\infty(G'^J)$ with*

   i) $\phi(\gamma g) = \phi(g)$ for all $\gamma \in \Gamma'^J$

   ii) $\phi(gr(\theta, \zeta)) = \phi(g)\zeta^m e^{ik\theta}$ for all $r(\theta, \zeta) \in K^J$

   iii) $\mathcal{L}_C\phi = \lambda^*\phi$

   iv) $\phi(g)$ is slowly increasing, meaning it is of polynomial growth in y.

Apparently the lifts of holomorphic Jacobi forms $f \in J_{k,m}$ are automorphic in this sense of type m, k, λ^* with $\lambda^* = (k - 1/2)(k - 5/2)/2$ and those of the skew-holomorphic forms $f \in J^*_{k,m}$ are of type $m, 1 - k, \lambda^*$ with the same λ^*. Another important example is given by the Eisenstein series

$$E_{k,m,\lambda}((\tau, z), s_1) = \sum_{\gamma \in \Gamma^J_N \backslash \Gamma'^J} (f_{k,m,\lambda,s_1}|_{k,m}[\gamma])(\tau, z) \tag{4.7}$$

with $\Gamma^J_N = \Gamma'^J \cap N^J$ and

$$f_{k,m,\lambda,s_1}(\tau, z) = e^m(\lambda^2\tau + 2\lambda z)y^{s_1-(k-1/2)/2}, \qquad \lambda \in \{0, \dots, 2m - 1\}.$$

This series is absolutely convergent for $\text{Re}(s_1) > 5/4$, and the vector

$$\mathbf{E}_{k,m} = (E_{k,m,\lambda})_\lambda$$

has a functional equation and analytic continuation by [Ar] Theorem 3.2. This $E_{k,m,\lambda}(\tau, z, s_1)$ may be lifted to a function on G'^J given by

$$E_{k,m,\lambda}(g, s) = \sum_{\gamma \in \Gamma^J_N \backslash \Gamma'^J} \phi_{m,s,0,k}((\lambda, 0, 0)\gamma g) \qquad \text{with } s = 2s_1 - 1$$

($\phi_{m,s,0,k}$ was defined in Proposition 3.2.9). This is an example for a form of type m, k, λ^* with $\lambda^* = (s^2 - 1)/2$.

4.2 The cusp condition for $G^J(\mathbb{R})$

In the general theory for a real reductive group G with discrete subgroup Γ, the discrete spectrum of the right regular representation ρ is separated from the rest by distinguishing as a **cuspidal part** $\mathcal{H}^0$ in $\mathcal{H} = L^2(\Gamma\backslash G)$ the closed subspace, invariant under ρ, spanned by the $\phi \in \mathcal{H}$ with

$$\phi^0_{N'}(g) := \int_{\Gamma^J \cap N'} \phi(n'g_0)dn' = 0 \quad \text{for almost all } g_0 \in G \text{ and all cuspidal } N'.$$

Here the notion "**cuspidal N'**" is as in Lang [La] p. 219-220 (and thus a bit different from, for instance, Harish-Chandra [HC] p. 8, where the adelic case is treated) and means N' is the unipotent radical of a $\mathbb{Q}$-parabolic $P' \subset G$ with

   $\Gamma \cap N'\backslash N'$ compact.

Now, besides the smallness of the center of $U(\mathfrak{g}_{\mathbb{C}}^J)$ already dealt with by localization in 3.1, it is one of the main features distinguishing our case from the
general theory of reductive groups, that this definition doesn't work here, as in
the direct generalization N' would come out as too big. But from the end of
1.3, we have a natural way to define here the notion of a cuspidal group and
a cusp in the following way (which then further down will be tied as usual to
the cusp condition iv) in the definition of a Jacobi form):

We take the maximal torus A of G''^J, a root decomposition belonging to A,
and get a two-parameter subgroup

$$N'^J = \left\{ \begin{pmatrix} 1 & x \\ 0 & 1 \end{pmatrix} (0, q, 1) =: n(x, q) : x, q \in \mathbb{R} \right\}$$

of G''^J belonging to the positive roots. Obviously,

$$(N'^J \cap \Gamma'^J) \backslash N'^J \simeq (\mathbb{Z} \backslash \mathbb{R})^2$$

is compact, and by [GGP] p. 95 we are led to introduce the following notions.

4.2.1 Definition. *A subgroup N^{*J} of G'^J is called* **horospherical** *if and only if
it is conjugate to N'^J, i.e. $N^{*J} = (N'^J)^g$, with $g \in G'^J(\mathbb{R})$, and* **cuspidal** *(for
Γ^J) if moreover*

$$(N^{*J} \cap \Gamma'^J) \backslash N^{*J} \quad \text{is compact.}$$

We easily have

4.2.2 Remark. A horospherical N^{*J} is cuspidal if and only if

$$g^{-1} \Gamma^J g \cap N'^J \text{ is a } \mathbb{Z}\text{-lattice of rank 2,}$$

and if and only if $(N^{*J})^\gamma$ is cuspidal for $\gamma \in \Gamma'^J$.

Moreover, by a longer but straightforward calculation we can prove:

4.2.3 Proposition. *A horospherical N^{*J} is cuspidal if and only if it is conjugate
to N'^J by an element of $G''^J(\mathbb{Q})$.*

As $B'^J = N'^J A Z'$ normalizes N'^J, the Γ^J–conjugacy classes $[(N'^J)^g]$ are
parametrized by the double cosets $[g]$ of $g \in G'^J$ in

$$\Gamma'^J \backslash G'^J / B'^J.$$

Using the proposition above, one is ready to call the classes $[g]$ with $g \in G'^J(\mathbb{Q})$
cusps for Γ'^J. But even independently from Proposition 4.2.3, we can easily
show the following characterization of cusps resp. cuspidal groups N^{*J}.

4.2.4 Proposition. *The Γ'^J–conjugacy classes of Γ^J–cuspidal subgroups of G'^J are in bijection to the set*

$$g_\lambda = (\lambda, 0, 1) \in H(\mathbb{R}) \qquad \text{with} \quad \lambda \in \mathbb{Q}/\mathbb{Z}.$$

Proof: i) By Remark 4.2.2 we have to look for $g = M(\lambda, \mu, 1)$ such that

$$(g^{-1}\Gamma'^J g) \cap N'^J \qquad \text{is a lattice of rank 2.}$$

From the SL(2)-theory (see for instance [La] p. 220-221) we know $M = 1$, and using the action of B'^J from the right we may restrict to g of type

$$g = g_\lambda = (\lambda, 0, 1), \qquad \lambda \in \mathbb{R}.$$

For $\gamma = M_0(p_0, q_0, 1) \in \Gamma'^J$ the condition

$$g_\lambda^{-1}\gamma g_\lambda \in N'^J \tag{4.8}$$

asks for $M_0 = n(x_0)$, $x_0 \in \mathbb{Z}$. Then we have

$$g_\lambda^{-1}n(x_0)(p_0, q_0, 1)g_\lambda$$
$$= n(x_0)\Big(p_0, \; q_0 - \lambda x_0, \; e(\lambda^2 x_0 - 2\lambda q_0 - \lambda p_0 x_0)\Big), \tag{4.9}$$

and (4.8) asks for $p_0 = 0$. We see that

$$g_\lambda^{-1}\Gamma^J g_\lambda \cap N'^J$$

is a $\mathbb{Z}$-lattice of rank 2 exactly if $e(\lambda^2 x_0 - 2\lambda q_0) = 1$ holds, i.e., if $\lambda \in \mathbb{Q}$, as x_0, q_0 are given as integers.

ii) The equality

$$\Gamma'^J g_\lambda B'^J = \Gamma'^J g_{\lambda'} B'^J$$

is equivalent to the existence of $g, \tilde{g} \in B'^J$ and $\gamma, \tilde{\gamma} \in \Gamma'^J$ with

$$\gamma g_\lambda g = \tilde{\gamma} g_{\lambda'} \tilde{g} \qquad \text{that is} \qquad g\tilde{g}^{-1} = g_{-\lambda}\gamma^{-1}\tilde{\gamma} g_{\lambda'}.$$

With

$$\hat{\gamma} := \gamma^{-1}\tilde{\gamma} = \begin{pmatrix} \hat{a} & \hat{b} \\ \hat{c} & \hat{d} \end{pmatrix}(\hat{p}, \hat{q}, 1)$$

the condition $g\tilde{g}^{-1} \in B'^J$ requires at first

$$\hat{c} = 0 \quad \text{and} \quad \hat{a} = \hat{d} = 1,$$

and then with

$$g_{-\lambda}\hat{\gamma}g_\lambda = \begin{pmatrix} 1 & \hat{b} \\ 0 & 1 \end{pmatrix}\Big(-\lambda + \lambda' + \hat{p}, \; \hat{q} - \hat{b}\lambda', \; e(-\lambda'(\hat{q} + \hat{p}\hat{b} - \hat{\lambda}\hat{b}))\Big)$$

finally

$$\lambda' - \lambda - \hat{p} = 0, \qquad \text{that is} \qquad \lambda \equiv \lambda' \bmod 1. \qquad \square$$

Now, we will use this to separate the discrete from the continuous part in $\mathcal{H}$.

4.2.5 Definition. *The **cuspidal part** $\mathcal{H}^0$ of the space $\mathcal{H} = L^2(\Gamma'^J \backslash G'^J)$ is distinguished by*

$$\mathcal{H}^0 = \left\{ \phi \in \mathcal{H} : \int_{(N^{*J} \cap \Gamma'^J) \backslash N^{*J}} \phi(ng_0) dn = 0 \quad \text{for almost all } g_0 \in G'^J \right.$$
$$\left. \text{and all cuspidal } N^{*J} \right\}.$$

Any Γ'^J–invariant function ϕ on G'^J for which the integral in the definition makes sense and which fulfills this "cuspidal condition" will be called **cuspidal**. For ϕ with $\phi(g\zeta) = \zeta^m \phi(g)$, $m \in \mathbb{Z}$, there is only a finite number of conditions to check.

4.2.6 Proposition. *ϕ is cuspidal exactly if for almost all $g_0 \in G'^J$ one of the following equivalent conditions holds.*

i) $\displaystyle\int_{(N^{*J} \cap \Gamma'^J) \backslash N^{*J}} \phi(ng_0) dn = 0 \qquad \text{for } N^{*J} = (N'^J)^{g_\lambda}$

with $g_\lambda = (\lambda, 0, 0)$, where

$$\lambda = \begin{cases} r/(2m), \ r = 0, \ldots, 2m - 1 \text{ s.t. } \lambda^2 m \in \mathbb{Z} & \text{if } m \neq 0, \\ 0 & \text{if } m = 0. \end{cases}$$

ii) $\displaystyle\int_{(N'^J \cap g_\lambda^{-1} \Gamma^J g_\lambda) \backslash N'^J} \phi(g_\lambda n g_0) \, dn = 0 \qquad \text{for } g_\lambda \text{ as above.}$

iii) $\displaystyle\int_{(N'^J \cap \Gamma^J) \backslash N'^J} \phi(ng_0) \bar{\psi}^{n,r}(n) \, dn = 0, \qquad \psi^{n,r}(n(x,q)) = e(nx + rq)$

     with

$$n, r \in \mathbb{Z} \quad \text{such that} \quad N = 4mn - r^2 = 0.$$

Proof: With

$$\Gamma_{N\lambda} = g_\lambda N'^J g_\lambda^{-1} \cap \Gamma'^J, \qquad N_{\Gamma\lambda} = N'^J \cap g_\lambda^{-1} \Gamma'^J g_\lambda,$$

we have the isomorphism

$$\Gamma_{N\lambda} \backslash N^{*J} \simeq N_{\Gamma\lambda} \backslash N'^J$$

(induced by conjugation with g_λ^{-1}), and the integral in the cuspidal condition may be written as

$$W_\lambda(g_0) \quad = \quad \int_{\Gamma_{N^\lambda}\backslash N^{*J}} \phi(ng_0)\,dn \quad = \quad \int_{N_{\Gamma^\lambda}\backslash N'^J} \phi(g_\lambda n g_\lambda^{-1} g_0)\,dn$$

$$= \int_{\mathcal{F}(N_{\Gamma^\lambda})} \phi(g_\lambda n(x,q) g_\lambda^{-1} g_0)\,dx\,dq,$$

where for a subgroup $\Gamma_0 \subset \Gamma'^J \cap N'^J$ a fundamental domain in N'^J is denoted by $\mathcal{F}(\Gamma_0)$. Using

$$g_\lambda n(x,q) g_\lambda^{-1} = n(x, q + \lambda x) e(\lambda^2 x + 2\lambda q),$$

we have

$$W_\lambda(g_0) = \int_{\mathcal{F}(N_{\Gamma^\lambda})} \phi(n(x, q + \lambda x) g_0) e^m (\lambda^2 x + 2\lambda q)\,dx\,dq.$$

The substitution $(x,q) \mapsto (x, q+\lambda x)$ amounts to changing $\mathcal{F}(N_{\Gamma^\lambda})$ into $\mathcal{F}(\Gamma_{N^\lambda})$, whence

$$W_\lambda(g_0) = \int_{\mathcal{F}(\Gamma_{N^\lambda})} \phi(n(x,q) g_0) e^m (2\lambda q - \lambda^2 x)\,dx\,dq.$$

The decomposition

$$\mathcal{F}(\Gamma_{N^\lambda}) = \bigcup_j \gamma_j \mathcal{F}(\Gamma'^J_N),$$

where $\gamma_j \in \Gamma'^J_N$ is a complete set of representatives for the finite abelian group $\Gamma_{N^\lambda}\backslash\Gamma'^J_N$, and the Γ'^J-invariance of ϕ lead to

$$W_\lambda(g_0) = \sum_j \chi_\lambda(\gamma_j) \int_{\mathcal{F}(\Gamma'^J_N)} \phi(n(x,q) g_0) \chi_\lambda(x,q)\,dx\,dq,$$

where χ_λ denotes the character of N'^J given by

$$\chi_\lambda(n(x,q)) = e^m(2\lambda q - \lambda^2 x).$$

The character sum $\sum_j \chi_\lambda(\gamma_j)$ for the finite group $\Gamma_{N^\lambda}\backslash\Gamma'^J_N$ is zero if the restriction of χ_λ to Γ'^J_N is not the trivial character. This is the case if and only if

$$2\lambda m \in \mathbb{Z} \qquad \text{and} \qquad \lambda^2 m \in \mathbb{Z}.$$

Thus the cusp condition $W_\lambda(g_0) = 0$ for almost all $g_0 \in G^J$ and all $\lambda \in \mathbb{Q}/\mathbb{Z}$ comes down to the finitely many cases denoted in the proposition. Condition iii) comes out remembering that for $4mn - r^2 = 0$ we have

$$e(nx + rq) \quad = \quad e^m\left(\left(\frac{r}{2m}\right)^2 x + 2\frac{r}{2m}q\right). \qquad \square$$

It is easy to relate the cusp conditions from the last proposition to the cusp condition for Jacobi forms. For a Γ'^J-invariant function ϕ on G'^J such that the integral exists, we define its (n, r)-**Whittaker-Fourier coefficient** by

$$W_\phi^{n,r}(g) = \int\limits_{(N'^J\cap\Gamma'^J)\backslash N'^J} \phi(ng)\bar{\psi}^{n,r}(n)\,dn. \tag{4.10}$$

4.2.7 Remark. For $\phi = \phi_f$ with

$$\phi_f(g) = j_{k,m}(g, (i, 0))f(\tau, z), \qquad f \in J_{k,m},$$

we have

$$W_\phi^{n,r}(g) = j_{k,m}(g, (i, 0))c(n, r)e(n\tau + rz).$$

This is straightforward, as we have $j_{k,m}(n, g(i, 0)) = 1$ and

$$W_\phi^{n,r}(g) = \int\limits_0^1\!\!\int\limits_0^1 j_{k,m}(ng, (i, 0))f(n(\tau, z))\bar{\psi}^{n,r}(n(x, q))\,dx\,dq$$

$$= j_{k,m}(g, (i, 0))\int\limits_0^1\!\!\int\limits_0^1 \sum c(n, r)e(n(\tau + x) + r(z + \xi)) \cdot e(-nx - r\xi)\,dx\,d\xi$$

$$= j_{k,m}(g, (i, 0))e(n\tau + rz)c(n, r).$$

4.2.8 Remark. For $\phi = \phi_f$ with

$$\phi_f(g) = j_{k,m}^*(g, (i, 0))f(\tau, z), \qquad f(\tau, z) \in J_{k,m}^*,$$

we have as well

$$W_\phi^{n,r}(g) = j_{k,m}^*(g, (i, 0))c(n, r)e(n\tau + rz + iy(r^2 - 4mn)/(2m)).$$

4.3   The discrete part and the duality theorem

We denote by $\mathcal{H}_m^0$ the closure of the subspace of $\mathcal{H} = L^2(\Gamma'^J\backslash G'^J)$ spanned by all $\phi \in \mathcal{H}$ with

$$\phi(g\zeta) = \zeta^m\phi(g) \quad \text{for all} \quad \zeta \in S^1,$$

and with the cusp condition

$$\int\limits_{(N'^J\cap\Gamma'^J)\backslash N'^J} \phi(g_\lambda \cap g_0)\,dn = 0 \quad \text{for almost all } g_0 \in G'^J \text{ and all } \lambda = r/(2m),$$

$$r = 0,\ldots,2m - 1 \text{ with } \lambda^2 m \in \mathbb{Z}.$$

As in the general theory (see for instance Godement [Go1] or Lang [La] p. 234) we have a discrete decomposition.

4.3.1 Theorem. *The representation ρ of G'^J given by right translation on $\mathcal{H}_m^0$ is completely reducible, and each irreducible component occurs only a finite number of times in it.*

As $\mathcal{H}$ has the ρ-invariant decomposition $\mathcal{H} = \oplus \mathcal{H}_m$, the same result holds for $\mathcal{H}^0 = \oplus \mathcal{H}_m^0$.

There is a **proof** of this theorem in [BeBö] 8 which follows the lines prescribed in Lang's book and Godement's article. We won't reproduce this proof entirely, but only indicate some steps, hoping that someone will find a more elegant proof.

I. For functions $\varphi \in \mathcal{C}_c^\infty(G'^J)$ one introduces an operator $T(\varphi)$ on $\mathcal{H}_m^0$ by

$$T(\varphi)\phi(g_1) = \int\limits_{G'^J} \phi(g_1 g_2)\varphi(g_2)\, dg_2 \quad \text{for} \quad \phi \in \mathcal{H}_m^0.$$

By general theorems ([La] p. 234) the assertion of the theorem follows if it is shown that there exists a number C_φ such that for all $\phi \in \mathcal{H}_m^0$

$$\|T(\varphi)\phi\| \le C_\varphi \|\phi\|_2$$

holds, where $\|\ \|$ ist the sup norm.

II. With the G'^J–biinvariant measure on G'^J given by

$$dg = y^{-2}\, dx\, dy\, d\theta\, dp\, dg\, \frac{d\zeta}{\zeta} \quad \text{for} \quad g = (n(x)t(y)r(\theta), p, q, \zeta)$$

we have by the periodicity of ϕ with $\Gamma_\infty^J = N'^J \cap \Gamma'^J$

$$T(\varphi)\phi(g_1) = \int\limits_{\Gamma_\infty^J \backslash G'^J} \sum_{\lambda,\mu \in \mathbb{Z}} \varphi(g_1^{-1} n(\lambda,\mu) g_2')\phi(g_2')\, dg_2'.$$

With

$$\varphi_{g_1,g_2}(\lambda,\mu) = \varphi(g_1^{-1} n(\lambda,\mu) g_2),$$

the kernel

$$K_\varphi(g_1,g_2) = \sum_{\varphi,\mu \in \mathbb{Z}} \varphi_{g_1,g_2}(\lambda,\mu)$$

may be expressed by the Poisson formula as

$$\begin{aligned}
K_\varphi(g_1,g_2) &= \sum_{\varphi,\mu \in \mathbb{Z}} \hat{\varphi}_{g_1,g_2}(\lambda,\mu) \quad (\hat{} = \text{Fourier transform}) \\
&= K_\varphi^0(g_1,g_2) + K_\varphi^1(g_1,g_2)
\end{aligned}$$

where

$$K_\varphi^0(g_1, g_2) = \sum_{4m\lambda + \mu^2 = 0} \hat{\varphi}_{g_1, g_2}(\lambda, \mu),$$

$$K_\varphi^1(g_1, g_2) = \sum_{4m\lambda + \mu^2 \neq 0} \hat{\varphi}_{g_1, g_2}(\lambda, \mu).$$

From now on we restrict to the case $m \neq 0$. The case $m = 0$ may be treated similarly (see [BeBö] p. 41).

III. By a routine calculation the cusp condition leads to

$$K_\varphi^0(g_1, g_2) = 0.$$

IV. We are left with

$$T(\varphi)\phi(g_1) = \int_{\Gamma_\infty^J \backslash G'^J / S^1} \sum_{4m\lambda + \mu^2 \neq 0} \int_{S^1} \hat{\varphi}_{g_1, g_2'', \zeta}(\lambda, \mu) \zeta_2^{m-1} \phi(g_2'') \, d\zeta_2 \, dg_2'',$$

where

$$\varphi_{g_1, g_2'', \zeta}(u, q) := \varphi(g_1^{-1} n(u, q) g_2'' \zeta).$$

V. For the **Siegel set**

$$\mathcal{S}(G'^J) = \{ g = (x, y, \theta, p, q, \zeta) : \quad 0 \leq x \leq 1, \; y \geq 1/2, \; \theta \in [0, 2\pi],$$
$$0 \leq p \leq 1/2, \; 0 \leq q \leq 1, \; \zeta \in S^1 \}$$

one has

$$G'^J = \Gamma'^J \mathcal{S}(G'^J).$$

We use the symbol Ω_G to denote a compact subset of a subgroup G of G'^J and abbreviate

$$\omega_{g_1} = g_1^{-1} t_1, \quad \omega_{g_1, g_2} = t_1^{-1} g_2 \quad \text{for} \quad g_i = n_i t_i r_i, \quad i = 1, 2,$$

to get

$$\varphi_{g_1, g_2'', \zeta}(u, q) = \varphi(\omega_{g_1} t_1^{-1} \tilde{n} t_1 \omega_{g_1, g_2''} \zeta_2).$$

Then, we can prove by some juggling around with compact sets:

Remark 1: There is a compact subset $\Omega_{G'^J}$ of G'^J such that for $g_1 \in \mathcal{S}(G^J)$ one has $\omega_{g_1} \in \Omega_{G'^J}$.

Remark 2: If $g_1 \in \mathcal{S}(G'^J)$ and $\varphi(g_1^{-1} n_\Gamma g_2) \neq 0$ for some $n_\Gamma \in \Gamma_\infty^J$, we may assume that modulo changes of g_2 on the left by an element of Γ_∞^J we have

$$\omega_{g_1, g_2} = t_1^{-1} g_2 \in \Omega_{G'^J}.$$

VI. Using this, we can easily modify the expression from part IV to get with $\zeta^* = \zeta_2 e(p_1^2 x - 2qp_1)$

$$T(\varphi)\phi(g_1) = \int\limits_{\Gamma_\infty^J \backslash G'^J / S_1} y_1^{3/2} \sum_{4m\lambda + \mu^2 \neq 0} \iint\limits_{S^1 \mathbb{R}^2} \varphi_{\omega_{g_1}^{-1}, \omega_{g_1, g_2'', \zeta^*}}(\tilde{x}, \tilde{q})$$

$$\cdot e((2mp_1 - \mu)y_1^{1/2}\tilde{q} + (mp_1^2 - (\lambda + \mu p_1))y_1\tilde{x})\, d\tilde{x}\, d\tilde{q}\, \zeta^{*m-1}\, d\zeta^*\, \phi(g_2'')\, dg_2''$$

In the inner integral there is a finite $\mathcal{C}^\infty$-function φ of $\tilde{x}$ and $\tilde{q}$ depending for $g_1 \in \mathcal{S}(G'^J)$ on parameters in compact sets (see remarks 1 and 2 in V.). Thus after d_1 partial integrations in $\tilde{q}$ and d_2 partial integrations in $\tilde{x}$ we obtain for this integral with

$$A = A_\mu = |2mp_1 - \mu|y_1^{1/2} \qquad \text{and} \qquad B = B_{\lambda,\mu} = |mp_1^2 - (\lambda + \mu p_1)|y_1$$

$$\left| \int\limits_{\mathbb{R}^2} \varphi_{\omega_{g_1}^{-1}, \omega_{g_1 g_2''}, \zeta^*}(\tilde{x}, \tilde{q}) e(A\tilde{q} + B\tilde{x})\, d\tilde{q}\, d\tilde{x} \right| \leq CA^{-d_1} B^{-d_2},$$

where C is a constant depending on φ, d_1, d_2 and $\mathcal{S}$. Here, at least, A or B has to be non–zero, because $A = B = 0$ is equivalent to $4m\lambda + \mu^2 = 0$. And this case is excluded in the sum, because it has been provided for with the cusp condition at the beginning. So, we have for $g_1 \in \mathcal{S}(G'^J)$

$$|T(\varphi)\phi(g_1)| \leq y_1^{3/2} \sum_{4m\lambda+\mu^2 \neq 0} C(\varphi, d_1, d_2, \mathcal{S})|A|^{-d_1}|B|^{-d_2} \int\limits_{\substack{\Gamma_\infty^J \backslash G'^J \\ g_2 \in t_1\Gamma_G}} |\phi(g_2)|\, dg_2.$$

By Schwarz's inequality, the last integral may be estimated by

$$\int |\phi(g_2)|\, dg_2 \leq (\mathrm{vol}\,(t_1\Omega_G))^{1/2}\|\phi\|_2.$$

Thus, the assertion of the theorem will be proved, if one can show that by chosing d_1, d_2 appropriately, there is an estimation of the series on the right hand side which is uniform in p_1, y_1 for $g_1 \in \mathcal{S}(G'^J)$, i.e., for $0 \leq p_1 \leq 1/2$, $y_1 \geq 1/2$.

VII. This estimation is given by the following

4.3.2 Lemma. One may choose $d_1 = d_1(p, \lambda, \mu)$ and $d_2 = d_2(p, \lambda, \mu)$ in such a way (depending on p, λ, μ) that one gets for fixed $m \neq 0$ a uniform bound for

$$R(y, p) := \sum_{4m\lambda+\mu^2 \neq 0} y^{3/2 - d_1/2 - d_2}|2mp - \mu|^{-d_1}|mp^2 - \mu p - \lambda|^{-d_2}$$

$((y, p)$ varying in a set of type $y \geq y_0 > 0$, p in a compact set$)$.

For a proof of this lemma we refer to [BeBö] p. 39–40 where the set of (λ, μ) is divided into three subsets and for each of these d_1 and d_2 are suitably chosen.

We will study the decomposition of $\mathcal{H}^0$ resp. $\mathcal{H}_m^0$ more closely. The discussion of the cusp condition has the following consequence.

4.3.3 Remark. We have for all $m, k \in N_0$

$$\mathcal{A}_{m,k}^0 \subset \mathcal{H}_m^0.$$

Proof: By Proposition 4.1.2 the function

$$|\phi_f(g)| = |f(\tau, z)| y^{k/2} e^{-2\pi m p^2 y}$$

is bounded for all $g \in {G'}^J$ and $k, m \in N_0$. Since a fundamental domain for the operation of ${\Gamma'}^J$ on ${G'}^J$ has finite measure with respect to the biinvariant dg, the assertion follows. $\square$

The discussion of the skew–holomorphic forms in Proposition 4.1.4 leads to the same result for $\mathcal{A}_{m,k}^{*0}$, defined as the subspace of bounded ϕ in $\mathcal{A}_{m,k}^*$.

As the lift ϕ_f of a Jacobi form $f \in J_{k,m}^{\mathrm{cusp}}$ is a lowest weight vector for the representation $\pi_{m,k}^+$ we expect here the same *duality relation* as in the SL(2)-theory between the dimension of the space of holomorphic cusp forms and the multiplicity of the representation in the right regular representation ρ:

4.3.4 Theorem (duality theorem). *For $m, k \in \mathbb{N}$ the multiplicity*

$$m_{m,k}^+ = \mathrm{mult}\,(\pi_{m,k}^+, \rho)$$

in the right regular representation ρ of ${G'}^J$ on $\mathcal{H}_m^0$ equals the dimension of the space of cusp forms of weight k and index m:

$$m_{m,k}^+ = \dim J_{k,m}^{\mathrm{cusp}}.$$

Proof: The finiteness of $m_{m,k}^+$ is already contained in Theorem 4.3.1.
i) Let

$$\mathcal{H}_m^0 = \bigoplus_n \mathcal{H}_{m,n}$$

be a decomposition of $\mathcal{H}_m$ into irreducible subspaces under the operation of right translation. In each $\mathcal{H}_{m,n}$ which is isomorphic to the space $\mathcal{H}_{\pi_{m,k}^+}$ of $\pi_{m,k}^+$ there exists (up to scalars) exactly one analytic lowest weight vector $\phi = \phi_{k,m}^n$ satisfying

$$\phi_{k,m}^n(gr(\theta, \zeta)) = \phi_{k,m}^n(g)\zeta^m e^{ik\theta} \qquad \text{for all} \quad \theta, \zeta.$$

By the discussion in 3.1 and 3.2 this is equivalent to

$$\mathcal{L}_{Z_0}\phi = 2\pi m\phi, \qquad \mathcal{L}_Z\phi = k\phi, \tag{4.11}$$

and

$$\mathcal{L}_{Y_-}\phi = \mathcal{L}_{X_-}\phi = 0. \tag{4.12}$$

From Proposition 4.1.2 we then know that ϕ is the lift of a cusp form, i.e.

$$m^*_{m,k} \leq \dim J^{cusp}_{k,m}.$$

ii) To obtain the other inequality, let f be an element of $J^{cups}_{k,m}$. Its lifting $\phi = \phi_f$ is a lowest weight vector for $\pi^+_{m,k}$ which fulfills the conditions just stated. From this we have that ϕ is an eigenfunction for the Casimir operator C from Section 3.1:

$$\mathcal{L}_C\phi = \lambda_k\phi, \qquad \lambda_k = k^2 - 3k + 5/4. \tag{4.13}$$

A given decomposition

$$\mathcal{H}^0_m = \underset{n}{\oplus} \mathcal{H}_{m,n} \qquad \text{induces} \qquad \phi = \sum_n \phi_n \quad \text{with} \quad \phi_n \in \mathcal{H}_{m,n}.$$

Since ϕ satisfies (4.13), each ϕ_n satisfies these relations too. We want to show that all ϕ_n needed in the decomposition of ϕ are vectors of lowest weight for a representation equivalent to $\pi^+_{m,k}$. This will be clear, if each $\mathcal{H}_{m,n}$ containing a component $\phi_n \neq 0$ is equivalent to the space $\mathcal{H}_{\pi^+_{m,k}}$. Since S^1 as a subgroup of G'^J commutes with G'^J, each $\rho(r(0,\varsigma))$ operates as a scalar on $\mathcal{H}_{m,n}$. The first equation in ii) of Proposition 4.1.2 shows that this scalar is ς^m, and this already fixes the type m of the representation belonging to $\mathcal{H}_{m,n}$. But in a representation where Z_0 operates as a scalar, C commutes with all representation operators, and, therefore, operates as a scalar too, say λ_n. Moreover by the same reasoning as in the discussion of the unitarizability we have for a vector space V with scalar product $\langle \, , \, \rangle$

$$\langle D_+\phi, \Psi \rangle = -\langle \phi, D_-\Psi \rangle \qquad \text{for} \quad \phi, \Psi \in V,$$

and hence

$$\langle D_+D_-\phi, \Psi \rangle = \langle \phi, D_+D_-\Psi \rangle$$

and thus the symmetry of

$$C = D_+D_- + D_-D_+ + (1/2)\Delta^2_1.$$

Then we have by (4.13) above for $V = \mathcal{H}$ and smooth functions

$$\lambda_n(\phi_n, \phi) = (\mathcal{L}_C\phi_n, \phi) = (\phi_n, \mathcal{L}_C\phi) = \lambda_k(\phi_n, \phi) \qquad \text{i.e.} \qquad \lambda_n = \lambda_k.$$

From the table of the possible eigenvalues of C in Proposition 3.1.10 we conclude $\pi = \pi^+_{m,k}$. □

Apparently, the considerations leading to the proof of the duality theorem show that Definition 4.1.8 indeed may be specialized to another characterization of Jacobi cusp forms.

4.3.5 Corollary. $J_{k,m}^{\mathrm{cusp}}$ *is isomorphic to the space of smooth functions* ϕ *on* G'^J *with*

   i) $\phi(\gamma g r(\theta, \zeta)) = \phi(g)\zeta^m e^{ik\theta}$ *for all* $\gamma \in \Gamma'^J$ *and* $r(\theta, \zeta) \in K^J$

   ii) $\mathcal{L}_C \phi = (k^2 - 3k + 5/4)\phi$

   iii) ϕ *is bounded (and ergo cuspidal).*

Moreover all this reasoning goes through also for the other types of representations to give analogous statements, for instance the duality theorem

$$m_{k,m}^- = \dim J_{k,m}^{*,\mathrm{cusp}}. \tag{4.14}$$

4.4   The continuous part

After the discussion of the discrete decomposition

$$\mathcal{H}_m^0 = \bigoplus_n \mathcal{H}_{m,n},$$

where the $\mathcal{H}_{m,n}$ are equivalent to the representation spaces of the representations $\pi_{m,k}^{\pm}, \pi_{m,s,\nu}$ discussed above, we now turn to the orthogonal complement $\mathcal{H}_m^c$ of $\mathcal{H}_m^0$ in $\mathcal{H}_m$ and expect a continuous decomposition as in the GL(2)-theory, described for instance in [Ge1] p. 161–162, [Ku2] p. 75 f, or [La] p. 239, i.e. something like

$$\mathcal{H}_m^c = \bigoplus_{\nu = \pm 1/2} \int_{\substack{\mathrm{Re}\, s = 0 \\ \mathrm{Im}\, s > 0}} \mathcal{H}_{m,s,\nu}\, ds, \qquad \mathcal{H}_{m,s,\nu} \text{ the space of } \pi_{m,s,\nu}.$$

To make this more precise, the starting point is to describe as usual $\mathcal{H}_m^c$ by *incomplete theta series* or as in [La] p. 240 by a *theta transform*. As already done in the last section, in the sequel we only give a sketch, following closely the exposition in [Be3], which we refer to for most of the proofs.

The theta transform and its adjoint

The space

$$\mathbf{Y} := N'^J \backslash G'^J$$

has the right G'^J–invariant measure

$$d\mu_{\mathbf{Y}} = y^{-2}\, dy\, d\theta\, dp\, d\zeta/\zeta.$$

Using this measure, we define
$$\mathcal{L}_m = L^2(N'^J \backslash G'^J)_m$$
as the space of functions φ on G'^J with
$$\varphi(\hat{n}g) = \zeta_1^m \varphi(g) \qquad \text{for } \hat{n} = n(x_1, q_1, \zeta_1) \in N^J Z$$
and
$$\int_Y |\varphi|^2 d\mu_\mathbf{Y} < \infty.$$

As above, we use
$$g_\lambda = (\lambda, 0, 1) \qquad \text{with } \lambda \in \Lambda,$$
where
$$\Lambda = \{\lambda = r/(2m) : r = 0, 1, \dots, 2m - 1\}.$$
Moreover, as in 4.2, we will use the subset
$$\Lambda^0 = \{\lambda \in \Lambda : \lambda^2 m \in \mathbb{Z}\}$$
and the group
$$\Gamma_N^J = \Gamma'^J \cap N'^J$$
with its conjugate subgroups
$$\Gamma_{N^\lambda} = \Gamma'^J \cap g_\lambda N'^J g_\lambda^{-1} \xrightarrow{\sim} N_{\Gamma^\lambda} = N'^J \cap g_\lambda^{-1} \Gamma'^J g_\lambda = (\Gamma_{N^\lambda})^{g_\lambda^{-1}}.$$
Here, and in several steps in the sequel, we work with the fundamental "commutation relation"
$$g_\lambda n(x, q) = n(x, q + \lambda x) g_\lambda e(2\lambda q + \lambda^2 x). \tag{4.15}$$
For appropriate functions $\varphi \in \mathcal{L}_m$ and $\lambda \in \Lambda$, we define the **theta transform** $\vartheta_\lambda \varphi$ by
$$\vartheta_\lambda \varphi(g) := \sum_{\gamma \in \Gamma_{N^\lambda} \backslash \Gamma'^J} \varphi(g_\lambda \gamma g).$$
This is well defined, because for $\gamma_0 = g_\gamma n g_\lambda^{-1} \in \Gamma_{N^\lambda}$ we have
$$\varphi(g_\lambda \gamma g) = \varphi(n g_\lambda \gamma g) = \varphi(g_\lambda (g_\lambda^{-1} n g_\lambda) \gamma g).$$
As to the convergence of $\vartheta_\lambda \varphi$ one has as in the usual SL(2)-theory (see for instance [La] p. 240)

- $\vartheta_\lambda \varphi$ is convergent for a Schwartz function φ.

- $\vartheta_\lambda \varphi$ has for fixed g only finitely many terms if supp(φ) is compact.

- $\vartheta_\lambda \varphi \in \mathcal{C}_c(\Gamma'^J \backslash G'^J)$ for $\varphi \in \mathcal{C}_c(N'^J \backslash G^J)_m$.

An **adjoint operator** ϑ_λ^* to ϑ_λ is given by an integral already used in the discussion of the cusp condition in Proposition 4.2.6. For $\phi \in \mathcal{H}_m$ and $\lambda \in \Lambda$ we define

$$\vartheta_\lambda^*\phi(g) := \int\limits_{N_{p\lambda}\backslash N'^J} \phi(g_\lambda^{-1}ng)\,dn.$$

4.4.1 Proposition. *Wherever the two operators ϑ_λ and ϑ_λ^* make sense, they are adjoint, i.e., we have for $\lambda \in \Lambda$*

$$\langle \vartheta_\lambda\varphi, \phi\rangle_{\mathcal{H}_m} = \langle \varphi, \vartheta_\lambda^*\varphi\rangle_{\mathcal{L}_m}.$$

Proof: From the definitions we get immediately

$$\langle \vartheta_\lambda\varphi, \phi\rangle_{\mathcal{H}_m} = \int\limits_{\Gamma'^J\backslash G'^J} \sum_{\Gamma_{N\lambda}\backslash\Gamma'^J} \varphi(g_\lambda\gamma g)\bar\phi(g)\,dg = \int\limits_{\Gamma_{N\lambda}\backslash G'^J} \varphi(g_\lambda g)\bar\phi(g)\,dg$$

and

$$\langle \varphi, \vartheta_\lambda^*\phi\rangle_{\mathcal{L}_m} = \int\limits_{N'^J\backslash G'^J} \varphi(\dot g) \int\limits_{N_{\Gamma\lambda}\backslash N^J} \bar\phi(g_\lambda^{-1}n\dot g)\,dn\,d\dot g$$

$$= \int\limits_{N_{\Gamma\lambda}\backslash G'^J} \varphi(g)\bar\phi(g_\lambda^{-1}g)\,dg.$$

Both expressions are equal, because the substitution $g_\lambda^{-1}g \mapsto g$ corresponds to the conjugation $N_{\Gamma\lambda} \mapsto \Gamma_{N\lambda}$. $\qquad\square$

4.4.2 Definition. *We denote by $\mathcal{H}_m^c$ the space of all incomplete theta series, i.e., the closure in $\mathcal{H}_m$ of the space spanned by all $\vartheta_\lambda\varphi$, $\varphi \in \mathcal{C}_c(N'^J\backslash G'^J)_m$ and $\lambda \in \Lambda$.*

4.4.3 Corollary. *$\mathcal{H}_m^0$ is the orthogonal complement of $\mathcal{H}_m^c$ in $\mathcal{H}_m$.*

Proof: For $\phi \in \mathcal{H}_m$

$$\langle \vartheta_\lambda\varphi, \phi\rangle_{\mathcal{H}_m} = 0 \qquad \text{for all } \varphi \in \mathcal{C}_c(N'^J\backslash G'^J)_m \text{ and } \lambda \in \Lambda$$

is equivalent to

$$\langle \varphi, \vartheta_\lambda^*\phi\rangle_{\mathcal{L}_m} = 0 \qquad\qquad \text{for all } \varphi \text{ and } \lambda.$$

As $\mathcal{C}_c(N'^J\backslash G'^J)_m$ is dense in $\mathcal{L}_m$, this leads to $\vartheta_\lambda^*\phi = 0$ for all λ, i.e., ϕ is cuspidal. $\qquad\square$

As in the cusp condition only the $\lambda \in \Lambda^0$ are essential, we are led to the following additional calculation.

4.4.4 Remark. For $\lambda \in \Lambda^0$ we may define

$$\vartheta_\lambda^\circ \varphi(g) = \sum_{\gamma \in \Gamma_N^J \backslash \Gamma'^J} \varphi(g_\lambda \gamma g), \tag{4.16}$$

and we have with $c_m := \sharp\{\Gamma_{N^\lambda} \backslash \Gamma_N^J\}$

$$\vartheta_\lambda \varphi(g) = c_m \vartheta_\lambda^\circ \varphi(g).$$

This comes out, as we here have $e^m(\pm 2q\lambda \pm x\lambda^2) = 1$ for all $q, x \in \mathbb{Z}$ and hence for $\gamma_0 = n(l, r)\ l, r \in \mathbb{Z}$

$$\varphi(g_\lambda \gamma_0 \gamma g) = \varphi(n(l, r + \lambda l)g_\lambda \gamma g)e^m(2r\lambda + l\lambda^2) = \varphi(g_\lambda \gamma g).$$

If $\gamma_j \in \Gamma_N^J$ $(j \in J)$ is a family representing $\Gamma_{N^\lambda} \backslash \Gamma_N^J$, we get

$$\vartheta_\lambda \varphi(g) = \sum_{\gamma_j \in \Gamma_{N^\lambda} \backslash \Gamma_N^J} \sum_{\gamma \in \Gamma_N^J \backslash \Gamma^J} \varphi(g_\lambda \gamma_j \gamma g) = c_m \vartheta_\lambda^\circ \varphi(g).$$

Similarly, we have for $\lambda \in \Lambda^0$ using the fundamental commutation relation (4.15) from above

$$\begin{aligned}
\vartheta_\lambda^* \phi(g) &= \int_{N_{\Gamma^\lambda} \backslash N'^J} \phi(g_\lambda^{-1} n g)\, dn \\
&= \int_{N_{\Gamma^\lambda} \backslash N'^J} \phi(n(x, q - \lambda x)g_\lambda^{-1} g)e^m(-2q\lambda + x\lambda^2)\, dn \\
&= \int_{N_{\Gamma^\lambda} \backslash N'^J} \phi(n(x, q)g_\lambda^{-1} g)e^m(-2q\lambda - x\lambda^2)\, dn.
\end{aligned}$$

Defining

$$\vartheta_\lambda^{\circ *} \phi(g) := \int_{\Gamma_N \backslash N'^J} \phi(n g_\lambda^{-1} g)\overline{\chi_\lambda(n)}\, dn, \quad \chi_\lambda(n) = e^m(2q\lambda + x\lambda^2)$$

we arrive at

$$\vartheta_\lambda^* \phi(g) = c_m \vartheta_\lambda^{\circ *} \phi(g),$$

and we can refine Proposition 4.4.1

4.4.5 Corollary. *Only the ϑ_λ° for $\lambda \in \Lambda^0$ are essential, and for these λ we have $\vartheta_\lambda^{\circ *}$ is adjoint to ϑ_λ°.*

The zeta transform and Eisenstein series

Now, we look for a relation between the elements of $\mathcal{H}_m^c$ and the spaces $\mathcal{H}_{m,s,\nu}$ of the principal series representations $\pi_{m,s,\nu}$. We recall from the discussion in 3.3 that the space $\mathcal{H}_{m,s}$ of functions ϕ on G^J with

$$\phi(b_0 g) = y_0^{(s+3/2)/2}\zeta_0\phi(g) \qquad \text{for } b_0 = n(x_0, q_0)t(y_0)\zeta_0 \in B^J$$

and

$$\|\phi\|^2 = \int\limits_{\mathcal{K}} |\phi(r(\theta, \hat{p}))|^2 \, d\theta \, d\hat{p} < \infty$$

is the sum of $\mathcal{H}_{m,s,1/2}$ and $\mathcal{H}_{m,s,-1/2}$, i.e., $\mathcal{H}_{m,s}$ may be spanned by the system (ϕ) of functions

$$\phi(g) = y^{(s+3/2)/2}\zeta^m e^m(pq + p^2 x)\varphi(\theta, py^{1/2})$$

with

$$\varphi(\theta, v) = e^{il\theta}e^{-v^2}H_j(v), \quad l \in \mathbb{Z}, j \in \mathbb{N}_0.$$

Again guided by the SL(2)-theory, we define a **zeta transform**

$$Z : \mathcal{L}_m \longrightarrow \mathcal{H}_{m,s}, \tag{4.17}$$

$$\varphi \longmapsto Z(\varphi, g, s) = \int\limits_0^\infty \varphi(t(y_0)g)y_0^{-s-1} \, dy_0.$$

As in [La] p. 243 this integral converges absolutely for $\varphi \in \mathcal{S}(N'^J\backslash G'^J)_m$ and $\sigma = \mathrm{Re}(s) > 0$, and is entire in s for $\mathcal{C}_c(N'^J\backslash G'^J)_m$. The standard formula

$$\int\limits_0^\infty \varphi(y_0 y)y_0^{-s-1} \, dy_0 = y^s\Gamma(l - s) \qquad \text{for } \varphi(y) = e^{-y}y^l$$

shows that the system (ϕ) of functions spanning $\mathcal{H}_{m,s}$ is up to factors the image of the zeta transform of

$$\varphi_{m,l,j}(g) = \zeta^m e^m(pz)e^{il\theta}e^{-y}H_j(v), \qquad v = (2\pi m)^{1/2}py^{1/2}.$$

Combining the zeta transform with the theta transform produces **Eisenstein series**

$$E_{m,\lambda}(\varphi, g, s) \quad := \quad \vartheta_\lambda^\circ Z(\varphi, g, s) \qquad (\lambda \in \Lambda^\circ) \tag{4.18}$$

$$= \quad \sum_{\gamma \in \Gamma_N^J\backslash\Gamma'^J} Z(\varphi, g_\lambda\gamma g, s).$$

These series converge absolutely for $\mathrm{Re}(s) > 3/2$ and $\varphi \in \mathcal{S}(N'^J\backslash G'^J)_m$ by the same reasoning which is used for the Eisenstein series discussed by Arakawa [Ar] (generalizing a notion by Eichler and Zagier [EZ], p. 17) and already mentioned in 4.1:

We put for $\lambda \in \Lambda^\circ$

$$f_{\kappa,m,\lambda,s_1}(\tau,z) = e^m(\lambda^2\tau + 2\lambda z)y^{s_1-\kappa}, \qquad \kappa = (k - 1/2)/2$$

and

$$E_{k,m,\lambda}((\tau,z),s_1) = \sum_{\gamma \in \Gamma_N^J \backslash \Gamma'^J} (f_{\kappa,m,\lambda,s_1}|_{k,m}[\gamma])(\tau,z).$$

This series is absolutely convergent for $\mathrm{Re}(s_1) > 5/4$, and if $k > 3$ and $s_1 = \kappa$ coincides with the holomorphic Eisenstein series of Eichler and Zagier. It is seen to be a special case of the Eisenstein series introduced above:

We may lift it to a function living on G'^J,

$$E_{k,m,\lambda}^A = \varphi_{k,m}(E_{k,m,\lambda}),$$

which may be recognized (using the notation from Proposition 3.3.1) as

$$E_{k,m,\lambda}^A(g,s_1) = \vartheta_\lambda^\circ \phi_{m,s,0,k}(g) \qquad \text{with } s_1 = (s+1)/2.$$

A bit more general, we put as well

$$E_{k,m,j,\lambda}(g,s') = \vartheta_\lambda^\circ \phi_{m,s,j,k}(g), \qquad s' = (s+1)/2 + 1/4,$$

and take these together as **column vectors**

$$\mathbf{E}_{m,k,j}(g,s) = (E_{\kappa,m,j,\lambda}(g,s))_{\lambda \in \Lambda_\kappa^\circ},$$

where Λ_k° corresponds bijectively to Arakawa's set R_k^{null} which he chose to get a set of linear independent series.

4.4.6 Remark. As the elements of the system (ϕ) spanning $\mathcal{H}_{m,s}$ are related by the ladder operators $\mathcal{L}_{X_\pm}$, $\mathcal{L}_{Y_\pm}$, we may in the same way reduce everything to a "cyclic" Eisenstein series $\mathbf{E}_{m,0,0}$. For instance we have

$$\mathbf{E}_{m,k,j} = \mathcal{L}_{Y_+}^j \mathbf{E}_{m,k,0}.$$

We have another kind of **reduction principle**.

4.4.7 Remark. As the space $\mathcal{L}_m$ may be spanned by functions of the type

$$\varphi(g) = \zeta^m e^m(pz)e^{ik\theta}H_j(v)\varphi_1(y), \qquad k \in \mathbb{Z}, j \in \mathbb{N}_0, \tag{4.19}$$

and we have, using the usual Mellin transform

$$L_{\varphi_1}(s) = \int_0^\infty \varphi_1(y)y^{-s-1}\,dy,$$

$$Z(\varphi,g,s) = L_{\varphi_1}(s)\phi_{m,s,j,k}(g),$$

we can say that everything concerning the column vector

$$\mathbf{E}_m(\varphi, g, s) = (E_{m,\lambda}(\varphi, g, s))_{\lambda \in \Lambda_k^\circ}, \quad \varphi \in \mathcal{L}_m$$

may be reduced to the *fundamental* vectors

$$L_{\varphi_1}(s)\mathbf{E}_{m,k,j}(g, s), \qquad k \in \mathbb{Z}, j \in \mathbb{N}_0,$$

and thus (by the previous remark) to Arakawa's Eisenstein series.

The functional equation for the Eisenstein series

In [Be3] p. 240 this reduction principle is used to prove the following statement by going back to a functional equation shown by Arakawa for his Eisenstein series by again going back to results about the metaplectic case.

4.4.8 Theorem. *For $\varphi \in \mathcal{L}_m^\circ$ and $\mathrm{Re}(s) = \sigma > 3/2$ one has a function $\hat{\varphi} \in \mathcal{L}_m^\circ$ such that the functional equation*

$$\mathbf{E}_m(\hat{\varphi}, g, s) = \Lambda(s)\mathbf{E}_m(\hat{\varphi}, g, 3/2 - s)$$

holds with a matrix $\Lambda(s)$ meromorphic in s with

$$\Lambda(s)\Lambda(3/2 - s) = 1.$$

Here $\mathcal{L}_m^\circ$ is a space of "nice" functions φ on G'^J, i.e., with

$$\varphi(\hat{n}g) = \zeta^m \varphi(g) \quad \text{for } \hat{n} \in N'^J S^1,$$
$$\varphi(-1g) = \varphi(g),$$
$$\varphi \in C^\infty \quad \text{and has compact support mod } N'^J$$

For details of the proof we refer here to the paper [Be3] but we would like to encourage further research to get a better proof, e.g. perhaps by coming down from results from the higher dimensional symplectic group.

The inner product formula

This formula is a first step in the direction of the aim to establish a Plancherel formula generalizing the one of the SL(2)-theory given in [Ku2] p. 85 or [La] p. 260, that is, to express the norm in $\mathcal{H}_m$ of a theta transform of φ by an integral over its "components" in the representation spaces $\mathcal{H}_{m,s}$. We recall that functions ϕ in $\mathcal{H}_{m,s}$ essentially only depend on the variables θ and $\hat{p} = py^{1/2}$, that is, they are fixed by their dependance on the space $\mathcal{K}$ with elements $\hat{r} = r(\theta, \hat{p})$ and measure $d\mu(\hat{r}) = d\hat{p}\, d\theta$.

4.4.9 Lemma. *For $\varphi, \psi \in \mathcal{L}_m^\circ$, $\lambda, \lambda' \in \Lambda_\kappa^\circ$ and $\mathrm{Re}\, s = \sigma > 5/4$ we have*

$$\langle \vartheta_\lambda^\circ \varphi, \vartheta_{\lambda'}^\circ \psi \rangle_{\mathcal{H}_m} = \frac{1}{2\pi i} \int_{\mathcal{K}} \int_{\mathrm{Re}\, s = \sigma} Z(\varphi_{\lambda,\lambda'}, \hat{r}, 3/2 - \bar{s})\overline{Z(\psi, \hat{r}, s)}\, ds\, d\mu(\hat{r})$$

Proof: Denoting the elements

$$\varphi_{\lambda,\lambda'} := \vartheta_{\lambda'}^{\circ *} \vartheta_{\lambda}^{\circ} \varphi$$

of the **constant term matrix** $\underline{\varphi}$, we get by the adjointness relation in the Proposition 4.4.1

$$\langle \vartheta_{\lambda}^{\circ}\varphi, \vartheta_{\lambda'}^{\circ}\psi \rangle_{\mathcal{H}_m} = \langle \varphi_{\lambda,\lambda'}, \psi \rangle_{\mathcal{L}_m}$$
$$= \int_{N'^J \backslash G'^J} \varphi_{\lambda,\lambda'}(\dot{g})\overline{\psi(\dot{g})}\, d\dot{g}$$

with

$$d\dot{g} = y^{-2}\, dy\, d\theta\, dp \qquad \text{for } \dot{g} = t(y)r(\theta, py^{1/2}).$$

Putting

$$f_1(y,\theta,\hat{p}) = \varphi_{\lambda,\lambda'}(t(y)r(\theta,\hat{p})) \text{ and } f_2(y,\theta,\hat{p}) = \psi(t(y)r(\theta,\hat{p}))$$

and using the formula (7.2.1.) from [Ku2] p. 78,

$$\int_0^\infty f_1(y)y^{-S_1} f_2(y)y^{-S_2}\frac{dy}{y} = \frac{1}{2\pi}\int_{-\infty}^{\infty} L_{f_1}(S_1 - it)\overline{L_{f_2}(S_2 - it)}\, dt,$$

with $S_1 = 3/2 - \sigma$ and $S_2 = \sigma$, we get

$$\langle \vartheta_{\lambda}^{\circ}\varphi, \vartheta_{\lambda'}^{\circ}\psi \rangle_{\mathcal{H}_m} = \int_{\mathcal{K}}\int_0^\infty f_1(y,\theta,\hat{p})y^{-3/2}\overline{f_2(y,\theta,\hat{p})}\,\frac{dy}{y}\, d\theta\, d\hat{p}$$
$$= \frac{1}{2\pi}\int_{\mathcal{K}}\int_{-\infty}^{\infty} L_{f_1}(3/2 - \sigma + it)\overline{L_{f_2}(\sigma + it)}\, dt\, d\theta\, d\hat{p}.$$

Resubstituting f_1 and f_2 and remembering

$$L_f(s) = \int_0^\infty f(y,\theta,\hat{p})y^{-s}\frac{dy}{y} = \int_0^\infty \varphi(t(y)\hat{r})y_0^{-s}\frac{dy_0}{y_0} = Z(\varphi,\hat{r},s)$$

we arrive at the assertion. $\qquad\qquad\qquad\qquad\qquad\qquad\qquad\qquad\Box$

The representations $\pi_{m,s,\nu}$ are unitary for $s \in i\mathbb{R}$. As $\mathcal{H}_{m,s}$ is spanned by functions ϕ with y–dependence given by $y^{s/2+3/4}$, this amounts here to the intention to shift the integration in the S–plane to the line $\operatorname{Re}(s) = \sigma = 3/4$. The reason that this may be done is the meromorphic continuation of the Eisenstein series following from its functional equation. This argument is also inherent in the following assertion whose full significance will become clear immediately in the proof of the main theorem.

4.4.10 Lemma. *For $\varphi, \psi \in \mathcal{L}_m^\circ$ and $\mathrm{Re}\, s = \sigma = 3/4$ we have*

$$\int\limits_{\mathcal{K}} Z(\varphi, \hat{r}, \bar{s})\overline{Z(\psi, \hat{r}, \bar{s})}\, d\mu(\hat{r})$$

$$= \int\limits_{\mathcal{K}} Z(\varphi, \hat{r}, s)\overline{Z(\hat{\psi}, \hat{r}, s)}^t \Lambda(s)\, d\mu(\hat{r}).$$

Proof: As in the preceding Lemma, we get for the inner product

$$\langle \vartheta_\lambda^\circ \varphi, E_{\lambda'}(\psi, \cdot, \bar{s})\rangle_{\mathcal{H}_m}$$

$$= \langle \varphi_{\lambda,\lambda'}, Z(\psi, \cdot, \bar{s})\rangle_{\mathcal{L}_m}$$

$$= \int\limits_0^\infty \int\limits_{\mathcal{K}} \varphi_{\lambda,\lambda'}(t(y)\hat{r})\overline{Z(\psi, t(y)\hat{r}, \bar{s})}\, d\mu(\hat{r}) y^{-5/2}\, dy$$

and by the transformation properly of Z "of type y^s"

$$= \int\limits_{\mathcal{K}} Z(\varphi_{\lambda,\lambda'}\hat{r}, 3/2 - s)Z(\psi, \hat{r}, \bar{s})\, d\mu(\hat{r}).$$

Substituting here $(\hat{\psi}, 3/2 - s)$ for (ψ, s) and multiplying with $\bar{\Lambda}(\bar{s})$ leads to

$$\langle \vartheta_\lambda^\circ \varphi, \sum \bar{\Lambda}_{\lambda',\mu'}(\bar{s})E_{\mu'}, (\hat{\psi}, \cdot, 3/2 - \bar{s})\rangle_{\mathcal{H}_m}$$

$$= \sum_{\mu'} \int\limits_{\mathcal{K}} Z(\varphi_{\lambda,\mu'}, \hat{r}, \bar{s})\overline{\bar{\Lambda}_{\lambda'\mu'}(\bar{s})}Z(\hat{\psi}, \hat{r}, s)\, d\mu(\hat{r}).$$

The left sides of both last equations are equal by the functional equation of the Eisenstein series. Thus, with $3/2 - s = \bar{s}$ for $\mathrm{Re}(s) = 3/4$ we get the assertion of the Lemma. $\qquad\square$

Now everything is ready for the final formula comparing inner products in $\mathcal{H}_m^c$ with those in the unitary representation spaces $\mathcal{H}_{m,s}$.

4.4.11 Theorem. *For $\varphi, \psi \in \mathcal{L}_m^\circ$ and $s = \sigma + it$ with $\sigma = 3/4$ we have the matrix equation*

$$((\langle \vartheta_\lambda^\circ \varphi, \vartheta_{\lambda'}^\circ \psi\rangle_{\mathcal{H}_m})_{\lambda,\lambda' \in \Lambda_k^\circ} = \frac{1}{2\pi} \int\limits_0^\infty \langle Z(\varphi, \cdot, s)\underline{\eta}^{-1}Z({}^t\psi, \cdot, s)\rangle_{\mathcal{H}_{m,s}}\, dt \tag{4.20}$$

with

$$\underline{\eta} = (\eta_{\lambda,\lambda'})_{\lambda,\lambda' \in \Lambda_k^\circ}, \qquad \eta_{\lambda,\lambda'} := \delta_{\lambda,\lambda'} + \delta_{1-\lambda,\lambda'}$$

Proof: By some further consideration we may take Lemma 4.4.9 and write

$$\langle \vartheta_\lambda^\circ \varphi, \vartheta_{\lambda'}^\circ \psi \rangle_{\mathcal{H}_m}$$

$$= \frac{1}{2\pi i} \int\limits_{\text{Re } s = \sigma} \int\limits_{\mathcal{K}} Z(\varphi_{\lambda,\lambda'}, \hat{r}, s)\overline{Z(\psi, \hat{r}, s)} \, d\mu(r) \, ds$$

$$= \frac{1}{2\pi} \int\limits_0^\infty \int\limits_{\mathcal{K}} Z(\varphi_{\lambda,\lambda'}, \hat{r}, \sigma + it)\overline{Z(\psi, \hat{r}, \sigma + it)} \, d\mu(r) \, dt$$

$$= \frac{1}{2\pi} \int\limits_0^\infty \int\limits_{\mathcal{K}} Z(\varphi_{\lambda,\lambda'}, \hat{r}, \sigma - it)\overline{Z(\psi, \hat{r}, \sigma - it)} \, d\mu(r) \, dt,$$

By Lemma 4.4.10 this amounts to

$$< \vartheta_\lambda^\circ \varphi, \vartheta_{\lambda'}^\circ \psi >_{\mathcal{H}_m}$$

$$= \frac{1}{2\pi} \int\limits_0^\infty \sum_{\mu'} \int\limits_{\mathcal{K}} Z(\varphi_{\lambda,\mu'}, \hat{r}, \sigma + it) \times$$

$$\times \overline{(\delta_{\mu',\lambda'} Z(\psi, \hat{r}, \sigma + it) + \Lambda_{\lambda',\mu'}(\sigma - it) Z(\hat{\psi}, \hat{r}, \sigma + it))} \, d\mu(\hat{r}) \, dt.$$

Now, the assertion follows from the matrix relation (E is here the unit matrix)

$$Z(\varphi, g, s)E + Z(\hat{\psi}, g, s)\Lambda(3/2 - s) = {}^t\underline{\eta}^{-1} Z({}^t\psi, g, s).$$

This relation is the result of some tedious calculation of the constant term matrices given in [Be3] Section 4.7. and 8.

5

Local Representations: The p-adic Case

In this chapter we turn to the local non-archimedean case and study the representation theory of $G^J(F)$, where F is a finite extension of some $\mathbb{Q}_p$. We will reach the goal of classifying all irreducible, admissible representations of $G^J(F)$ by using the fundamental relation $\pi \simeq \tilde{\pi} \otimes \pi_{SW}^m$ and the classification of representations of the metaplectic group given by Waldspurger in [Wa1].

Roughly speaking, all non-supercuspidal irreducible representations of Mp are obtained by parabolic induction. This result can be taken over to the Jacobi group by making the isomorphism $\pi \simeq \tilde{\pi} \otimes \pi_{SW}^m$ explicit in the context of standard models for induced representations (Theorem 5.4.2).

The importance of this explicit isomorphism will also become apparent while discussing intertwining operators in Section 5.6. We will see there how the naturally defined intertwining integrals correspond on both sides, and how the analytic continuation can be established. The discussion of Whittaker models in Section 5.7 will fill the last gap for a complete classification of irreducible, admissible representations. Finally, for global applications, it is important to single out the unitary representations.

5.1 Smooth and admissible representations

Let F be a p-adic field, i.e., a finite extension of some $\mathbb{Q}_p$, where the prime number p is the characteristic of the residue field of F. Let further be

- $q = p^f$ the number of elements of the residue field.
- $\mathcal{O}$ the ring of integers of F.
- $\mathfrak{p}$ the maximal ideal of $\mathcal{O}$.

- $v = v_F$ the normalized valuation of F.
- $\omega \in F$ an element with $v(\omega) = 1$.

This section deals with the Jacobi group over F. So let G^J, G, H, ... be the F-rational points of the corresponding algebraic groups (as before $G = SL(2)$). We stick to the usual notions for representations of p-adic groups. In particular, we call a representation of G^J on some complex vector space V **smooth** if V is covered by the subspaces

$$V^K = \{v \in V : gv = v \text{ for all } g \in K\}$$

where K runs through the open and compact subgroups of G^J. If, moreover, every V^K is finite dimensional, we call the representation **admissible**. More details on the general representation theory of p-adic groups can for instance be found in [Ca].

In Section 2.6 the fundamental fact was already stated that every smooth representation π of G^J with non-trivial central character is of the form

$$\pi = \tilde{\pi} \otimes \pi_{SW}^m$$

with $m \in F^*$ and $\tilde{\pi}$ a smooth representation of Mp. The standard model for both the Schrödinger and the Weil representation is the Schwartz space $\mathcal{S}(F)$ of locally constant compactly supported functions $F \to \mathbb{C}$. The explicit formulas for these representations were already given in Sections 2.2 resp. 2.5, but we recall them here to have them at hand:

$$\left(\pi_s^m(\lambda, \mu, \kappa)f\right)(x) = \psi\Big(m(\kappa + (2x + \lambda)\mu)\Big)f(x + \lambda).$$

$$\left(\pi_W^m\begin{pmatrix} 1 & b \\ 0 & 1 \end{pmatrix}f\right)(x) = \psi^m(bx^2)f(x).$$

$$\left(\pi_W^m\begin{pmatrix} a & 0 \\ 0 & a^{-1} \end{pmatrix}f\right)(x) = \delta_m(a)|a|^{1/2}f(ax).$$

$$\left(\pi_W^m\begin{pmatrix} 0 & 1 \\ -1 & 0 \end{pmatrix}f\right)(x) = \gamma(1)\hat{f}(x) = \gamma_m(1)\int_F \psi(2mxy)f(y)\,dy.$$

(Throughout the chapter ψ denotes the additive standard character defined in Section 2.2.) Here we abbreviate

$$\delta_m(a) = (a, -1)\gamma(a)\gamma(1)^{-1}.$$

This function is called the **Weil character** and will be discussed more closely in Section 5.3. The Weil constant $\gamma = \gamma_m$ was introduced in Section 2.5. We will mainly be concerned with admissible representations, for which there are the following results.

5.1.1 Lemma. *The Schrödinger and the Schrödinger-Weil representations are admissible.*

Proof: The second assertion certainly follows from the first. Let K be an open compact subgroup of H. It is clear that K contains a set of the form

$$\omega^\alpha \mathcal{O} \times \omega^\beta \mathcal{O} \times \omega^\gamma \mathcal{O} \qquad \text{with integers } \alpha, \beta, \gamma.$$

The invariance of $f \in \mathcal{S}(F)$ under $(\lambda, 0, 0)$ with $\lambda \in \omega^\alpha \mathcal{O}$ means that f is invariant under additive translations by $\omega^\alpha \mathcal{O}$. The invariance under $(0, \mu, 0)$ with $\mu \in \omega^\beta \mathcal{O}$ easily implies that f has support in $(2m\omega^\beta)^{-1}\mathcal{O}$. These two facts together show that $\mathcal{S}(F)^K$ is finite-dimensional. $\qquad\square$

5.1.2 Proposition. *π is admissible if and only if $\tilde{\pi}$ is admissible.*

Proof: Assume $\tilde{\pi} : \mathrm{Mp} \to \mathrm{GL}(V)$, also regarded as a projective representation of $\mathrm{SL}(2)$, is admissible and let $K \subset G^J$ be a compact open subgroup. Given $\Phi \in V \otimes \mathcal{S}(F)^K$ we can write it as

$$\Phi = \sum_i \varphi_i \otimes f_i$$

with linearly independent $\varphi_i \in V$ and linearly independent $f_i \in \mathcal{S}(F)$ (simply take the lowest possible number of summands). The intersection $K_1 := K \cap H$ is a compact open subgroup of H. From

$$\Phi = \pi(k_1)\Phi = \sum_i \varphi_i \otimes \pi_s^m(k_1)f_i \qquad \text{for all } k_1 \in K_1$$

and the linear independence of the φ_i it follows that $f_i \in \mathcal{S}(F)^{K_1}$ for each i. Choose a small enough compact open subgroup K_0 of $K \cap G$ such that $\pi_w^m(K_0)$ fixes each element of the finite-dimensional space $\mathcal{S}(F)^{K_1}$ (Lemma 5.1.1). Then from $\pi(k_0)\Phi = \Phi$ for all $k_0 \in K_0$ and the linear independence of the f_i it follows that $\varphi_i \in V^{K_0}$. What we have shown is that

$$(V \otimes \mathcal{S}(F))^K \subset V^{K_0} \otimes \mathcal{S}(F)^{K_1},$$

and the admissibility of π follows from that of $\tilde{\pi}$. It is easy to see that the converse is true. $\qquad\square$

5.2 Whittaker models for the Schrödinger-Weil representation

In this section we establish the existence and uniqueness of Whittaker models for the Schrödinger and the Schrödinger-Weil representations. As a commutative subgroup of H we take

$$N_H = \{(0, \mu, 0) : \mu \in F\} \simeq F,$$

and we are at first interested in Whittaker models for π_S^m with respect to this subgroup and the character ψ^r of N_H, where $r \in F$. That is, we are looking for a space of locally constant functions $W : H \to \mathbb{C}$ with the property

$$W((0, \mu, 0)h) = \psi^r(\mu)W(h) \qquad \text{for all } \mu \in F, \, h \in H,$$

such that right translation on this space defines a representation of H equivalent to π_S^m. The existence and uniqueness of such a space is equivalent to the fact that the space of linear functionals $l : \mathcal{S}(f) \to \mathbb{C}$ with the property

$$l\left(\pi_S^m(0, \mu, 0)f\right) = \psi^r(\mu)f \qquad \text{for all } \mu \in F, \, f \in \mathcal{S}(F),$$

is one-dimensional. Such a functional is called a $\psi^{m,r}$-Whittaker functional, and the associated Whittaker model shall be denoted $\mathcal{W}_S^{m,r}$.

5.2.1 Theorem. *For any $r \in F$ there exists a unique Whittaker model $\mathcal{W}_S^{m,r}$ for π_S^m. The associated Whittaker functional is given by*

$$l_S : \mathcal{S}(F) \longrightarrow \mathbb{C},$$
$$f \longmapsto f\left(\frac{r}{2m}\right).$$

The proof is taken from [Ho], which in turn goes back to [Wa1] and [Be6]. First we need a lemma.

5.2.2 Lemma. *For a functional $l : \mathcal{S}(F) \to \mathbb{C}$ and a point $\xi \in F$ the following statements are equivalent.*

   i) *There exists $c \in \mathbb{C}$ such that $l(f) = cf(\xi)$ for all $f \in \mathcal{S}(F)$.*

   ii) *$l(f) = 0$ for all $f \in F$ with $f(\xi) = 0$.*

Proof: Certainly ii) follows from i). Conversely, assume $f(\xi) = 0$ implies $l(f) = 0$. Let θ be the characteristic function of $\xi + \mathcal{O}$. Then for arbitrary $f \in \mathcal{S}(F)$ the function $f - f(\xi)\theta$ vanishes at ξ, hence

$$0 = l(f - f(\xi)\theta) = l(f) - l(\theta)f(\xi).$$

So i) follows with $c = l(\theta)$. $\qquad\qquad\qquad\qquad\qquad\qquad\qquad\qquad\quad \square$

Proof of Theorem 5.2.1: A very simple calculation shows that l_S is indeed a non-trivial ψ^r-Whittaker functional. It remains to show that l_S is unique up to scalars. So let l be another functional of this type. By the lemma, it is enough to show that $l(f) = 0$ for $f \in \mathcal{S}(F)$ with $f(r/2m) = 0$. Now, by Fourier inversion, there exists $\hat{f} \in \mathcal{S}(F)$ such that

$$f(x) = \int_F \hat{f}(y)\psi(xy) \, dy.$$

Since $\hat{f}$ is locally constant with compact support this may be written as a finite sum:

$$f(x) = \sum_{y \in \omega^a \mathcal{O} / \omega^b \mathcal{O}} \hat{f}(y) \psi(xy)$$

with suitable $a, b \in \mathbb{Z}$, $a < b$. We multiply this by the characteristic function $\mathbf{1}_{\omega^k \mathcal{O}}$ of $\omega^k \mathcal{O}$, where k is chosen small enough such that $\omega^k \mathcal{O}$ contains the support of f, and obtain

$$f = \sum_{y \in \omega^a \mathcal{O} / \omega^b \mathcal{O}} \hat{f}(y) \pi_s^m \left(0, \frac{y}{2m}, 0\right) \mathbf{1}_{\omega^k \mathcal{O}}.$$

Since $f(r/2m) = 0$ this can also be written as

$$f = \sum_{y \in \omega^a \mathcal{O} / \omega^b \mathcal{O}} \hat{f}(y) \left(\pi_s^m \left(0, \frac{y}{2m}, 0\right) \mathbf{1}_{\omega^k \mathcal{O}} - \psi\left(\frac{ry}{2m}\right) \mathbf{1}_{\omega^k \mathcal{O}}\right).$$

Application of our Whittaker functional l on both sides yields $l(f) = 0$, and we are done. $\qquad\square$

5.2.3 Corollary. *For every $M \in G$ there is exactly one non-trivial space $\mathcal{W}_M$ of locally constant functions $f : H(F) \to \mathbb{C}$ with the following properties:*

i) *$\mathcal{W}_M$ is stable under right translations.*

ii) *The representation on $\mathcal{W}_M$ defined by right translation is isomorphic to π_s^m.*

iii) *Every $f \in \mathcal{W}_M$ satisfies*

$$f\Big(\big((0, \mu)M, \kappa\big)h\Big) = \psi^m(\kappa) f(h) \qquad \text{for all } \mu, \kappa \in F, \ h \in H.$$

Proof: For $M = 1$ this is just the theorem with $r = 0$. Let $L(M)$ be the space of linear functionals $l : \mathcal{S}(F) \to \mathbb{C}$ such that

$$l\Big(\pi_s^m((0, \mu)M, 0)f\Big) = l(f) \qquad \text{for all } f \in \mathcal{S}(F), \ \mu \in F.$$

Then, very similar to usual Whittaker models, the assertion is equivalent with the fact that $\dim(L(M)) = 1$. If we extend π_s^m on $\mathcal{S}(F)$ to the Schrödinger-Weil representation π_{SW}^m, then it is easy to check that the linear map

$$L(1) \longrightarrow L(M),$$
$$l \longmapsto \Big(f \mapsto l(\pi_{SW}^m(M)f)\Big)$$

is an isomorphism of vector spaces (use (2.3) in Section 2.5). By the case $M = 1$ already known, the proof is complete. $\qquad\square$

Now we discuss Whittaker models for the Schrödinger-Weil representation. The subgroup under which we want to have the Whittaker transformation property is as in the real case

$$N^J = \left\{ \begin{pmatrix} 1 & x \\ 0 & 1 \end{pmatrix} (0, \mu, 0) : x, \mu \in F \right\} \simeq F^2.$$

But we have to be a bit careful because π_{sw}^m is projective with cocycle λ. So what we are looking for is a space of locally constant functions $W : G^J \to \mathbb{C}$ with the property

$$W \left(\begin{pmatrix} 1 & x \\ 0 & 1 \end{pmatrix} (0, \mu, 0) M h \right) = \lambda \left(\begin{pmatrix} 1 & x \\ 0 & 1 \end{pmatrix}, M \right) \psi^\nu(x) \psi^r(\mu) W(Mh)$$

for all $x, \mu \in F$, $M \in G$, $h \in H$, where ν and r are elements of F, and such that the following operation ρ of G^J defines a projective representation equivalent to π_{sw}^m:

$$(\rho(g')W)(g) = \lambda(g, g')W(gg').$$

If such a space exists, we denote it by $\mathcal{W}_{sw}^{m,\nu,r}$. It is again easy to see that the existence and uniqueness of $\mathcal{W}_{sw}^{m,\nu,r}$ is equivalent with the existence and uniqueness (up to scalars) of a $\psi^{\nu,r}$-Whittaker functional, i.e. a linear map $l : \mathcal{S}(F) \to \mathbb{C}$ with

$$l \left(\pi_{sw}^m \left(\begin{pmatrix} 1 & x \\ 0 & 1 \end{pmatrix} (0, \mu, 0) \right) f \right) = \psi^\nu(x) \psi^r(\mu) l(f) \qquad \text{for } x, \mu \in F, \ f \in \mathcal{S}(F).$$

5.2.4 Theorem. *A Whittaker model $\mathcal{W}_{sw}^{m,\nu,r}$ for the representation π_{sw}^m exists if and only if $\nu = r^2/4m$. In this case it is unique, and the Whittaker functional l_{sw} is given by evaluating functions at $r/2m$:*

$$l_{sw}(f) = f \left(\frac{r}{2m} \right) \qquad \text{for all } f \in \mathcal{S}(F).$$

Because ν is determined by r, we denote $\mathcal{W}_{sw}^{m,\nu,r}$ simply by $\mathcal{W}_{sw}^{m,r}$.

Proof: If a $\psi^{\nu,r}$-Whittaker functional l_{sw} for π_{sw}^m is given, then by the definitions it is also a ψ^r-Whittaker functional for π_s^m. This proves the uniqueness in view of Theorem 5.2.1. The rest is easy: If we take l_s for l_{sw}, then

$$l_{sw} \left(\pi_{sw}^m \left(\begin{pmatrix} 1 & x \\ 0 & 1 \end{pmatrix} (0, \mu, 0) \right) f \right) = \left(\pi_w^m \begin{pmatrix} 1 & x \\ 0 & 1 \end{pmatrix} \pi_s^m(0, \mu, 0) f \right) \left(\frac{r}{2m} \right)$$

$$= \psi^m \left(\frac{xr^2}{4m^2} \right) \psi^m \left(\frac{2\mu r}{2m} \right) f \left(\frac{r}{2m} \right)$$

$$= \psi \left(\frac{r^2}{4m} x + r\mu \right) l_{sw}(f).$$

So l_{sw} is a $\psi^{\nu,r}$-Whittaker functional exactly for $\nu = r^2/4m$. $\square$

5.2.5 Corollary. *The Weil representation π_W^m has a ψ^ν-Whittaker model if $\nu \in mF^{*2}$.*

We pay special attention to the Whittaker model $\mathcal{W}_{SW}^{m,0}$, which exists by the theorem. Using the known properties of the cocycle λ from [Ge2], one easily establishes for any W in this model the transformation formula

$$W\left(\begin{pmatrix} a & x \\ 0 & a^{-1} \end{pmatrix}(0, \mu, \kappa)Mh\right)$$
$$= \lambda\left(\begin{pmatrix} a & 0 \\ 0 & a^{-1} \end{pmatrix}, M\right)\delta_m(a)\psi^m(\kappa)|a|^{1/2}W(Mh), \qquad (5.1)$$

where $a \in F^*$, $x, \mu, \kappa \in F$, $M \in G$ and $h \in H$. If we extend W to a function $\widetilde{W}$ on $\widetilde{G^J}$ by setting

$$\widetilde{W}(g, \varepsilon) = \varepsilon W(g),$$

then this transformation property goes over to

$$\widetilde{W}\left(\left(\begin{pmatrix} a & x \\ 0 & a^{-1} \end{pmatrix}(0, \mu, \kappa), \varepsilon\right)g\right) = \varepsilon\delta_m(a)\psi^m(\kappa)|a|^{1/2}\widetilde{W}(g). \qquad (5.2)$$

This may be interpreted as follows. Consider the subgroup

$$\widetilde{B^J} = \left\{\left(\begin{pmatrix} a & x \\ 0 & a^{-1} \end{pmatrix}(0, \mu, \kappa), \varepsilon\right) : a \in F^*, \ x, \mu, \kappa \in F, \ \varepsilon \in \{\pm 1\}\right\} \qquad (5.3)$$

of $\widetilde{G^J}$ and its character

$$\chi\left(\left(\begin{pmatrix} a & x \\ 0 & a^{-1} \end{pmatrix}(0, \mu, \kappa), \varepsilon\right) = \varepsilon\delta_m(a)\psi^m(\kappa). \qquad (5.4)$$

Then the space of the induced representation

$$\text{ind}_{\widetilde{B^J}}^{\widetilde{G^J}}\chi$$

consists exactly of functions $\widetilde{W} : \widetilde{G^J} \to \mathbb{C}$ satisfying (5.2) and invariant on the right by an open subgroup of $\widetilde{G^J}$. The latter condition is also satisfied by the above $\widetilde{W}$ coming from $W \in \mathcal{W}_{SW}^{m,0}$, because π_{SW}^m is smooth. We see that $W \mapsto \widetilde{W}$ maps $\mathcal{W}_{SW}^{m,0}$ into $\text{ind}_{\widetilde{B^J}}^{\widetilde{G^J}}\chi$. If we let $\widetilde{G^J}$ act on $\mathcal{W}_{SW}^{m,0}$ by

$$(\rho(g', \varepsilon)W)(g) = \varepsilon\lambda(g, g')W(gg'),$$

i.e. we view π_{SW}^m as a representation of $\widetilde{G^J}$, then $W \mapsto \widetilde{W}$ becomes an intertwining map. Since π_{SW}^m is irreducible, we have proved the following.

5.2.6 Proposition. *The Schrödinger-Weil representation π_{SW}^m of $\widetilde{G^J}$ occurs as a subrepresentation in the induced representation*

$$\text{ind}_{\widetilde{B^J}}^{\widetilde{G^J}}\chi,$$

where $\widetilde{B^J}$ and the character χ are given by (5.3) and (5.4).

5.3   Representations of the metaplectic group

In this section we repeat the classification of the irreducible, admissible representations of the metaplectic group Mp over the p-adic field F, as was given by Waldspurger in [Wa1]. We are interested in the *genuine* representations, i.e., those which do not factor through $G = \mathrm{SL}(2, F)$, or equivalently, those for which $(1, -1) \in$ Mp operates as the negative of the identity. At first we have the following lemma ([Wa1] II, lemme 3).

5.3.1 Lemma. *Every genuine irreducible admissible representation of* Mp *is infinite dimensional.*

Next, one can single out the *supercuspidal* representations. An irreducible admissible representation $\pi :$ Mp $\to \mathrm{GL}(V)$ is called **supercuspidal** if for every $v \in V$ there is an integer n such that

$$\int\limits_{\omega^n \mathcal{O}} \pi \begin{pmatrix} 1 & x \\ 0 & 1 \end{pmatrix} v \, dx = 0.$$

Here one should remember that the short exact sequence defining the metaplectic group splits over N, so that N can be considered a subgroup of Mp, as we did. By a familiar reasoning (cf. [Ge2] p. 96) one comes to the conclusion that any non-supercuspidal irreducible admissible representation is a subquotient of some representation induced by a character of the 'torus'

$$\tilde{A} = \left\{ \left(\begin{pmatrix} a & 0 \\ 0 & a^{-1} \end{pmatrix}, \varepsilon \right) : a \in F^*,\ \varepsilon \in \{\pm 1\} \right\}.$$

We are interested in genuine characters of $\tilde{A}$, i.e., those which map $(1, -1) \in$ Mp to -1. One of them is what we call the **Weil character** δ_u. It is defined by

$$\delta_u \left(\begin{pmatrix} a & 0 \\ 0 & a^{-1} \end{pmatrix}, \varepsilon \right) = \varepsilon(a, -1)\gamma_u(a)\gamma_u(1)^{-1}$$

where $(\ ,\ )$ is the Hilbert symbol and γ_u the Weil constant, both explained in Section 2.5. The parameter $u \in F$ will later on usually be fixed to $-m$ when dealing with representations of G^J with central character ψ^m. We write $\delta_u(a)$ for $\delta_u(d(a), 1)$. The following lemma gives the basic properties of the Weil character. Proofs can be found in [Rao] and [Sch1] (see also [Wa1], lemme 1, and [GePS2], p. 150).

5.3.2 Lemma.

  i) $\delta_u(a)\delta_u(a') = (a, a')\delta_u(aa')$ *for all* $a, a' \in F^*$. *In other words,* δ_u *is a (genuine) character of* $\tilde{A}$.

  ii) $\delta_u(a)^{-1} = (a, -1)\delta_u(a) = \delta_{-u}(a)$ *for all* $a \in F^*$.

  iii) $\delta_u(a) = 1$ *for all* $a \in F^{*2}$.

If the residue characteristic of F is not 2, then we have in addition:

iv) $\delta_{ur}(a) = (a,r)\delta_u(a)$ for all $a, r \in F^*$.

v) $\delta_u(a) = 1$ for all $a \in \mathcal{O}^*$, if $v(u)$ is even.

vi) $\delta_u(a) = (a,u)$ for all $a \in \mathcal{O}^*$.

Other genuine characters of $\tilde{A}$ can be built simply by adjoining characters of A: If χ is a character of A, then

$$\left(\begin{pmatrix} a & 0 \\ 0 & a^{-1} \end{pmatrix}, \varepsilon \right) \longmapsto \varepsilon \delta_u(a)\chi(a)$$

defines a genuine character of $\tilde{A}$. These are the ones to be used now for the induction procedure. Namely let $\mathcal{B}_{\chi,u}$ be the space of functions $\varphi : \mathrm{Mp} \to \mathbb{C}$ with the following two properties:

i) φ is right invariant by some open subgroup of Mp.

ii) For every $a \in F^*$, $\varepsilon \in \{\pm 1\}$, $x \in F$ and $g \in \mathrm{Mp}$ the following holds:

$$\varphi \left(\left(\begin{pmatrix} a & 0 \\ 0 & a^{-1} \end{pmatrix}, \varepsilon \right) \begin{pmatrix} 1 & x \\ 0 & 1 \end{pmatrix} g \right) = \varepsilon \delta_u(a)|a|\chi(a)\varphi(g).$$

The action of Mp on $\mathcal{B}_{\chi,u}$ is given by right translation and denoted by $\rho_{\chi,u}$. Here is Waldspurger's result concerning the irreducibility of these induced representations:

5.3.3 Theorem.

i) If $\chi^2 \neq |\,|$ and $\chi^2 \neq |\,|^{-1}$, then $\rho_{\chi,u}$ is irreducible.

ii) If $\chi^2 = |\,|$, then one can find $\xi \in F^*$ such that $\chi = |\,|^{1/2}(\cdot, \xi)$, where $(\,,\,)$ is the Hilbert symbol. $\mathcal{B}_{\chi,u}$ contains exactly one non-trivial proper invariant subspace $\mathcal{B}_{\xi,u}$. The representation on the factor space is isomorphic to the positive (even) Weil representation with character $\psi^{u\xi}$.

iii) If $\chi^2 = |\,|^{-1}$, then $\chi = |\,|^{-1/2}(\cdot, \xi)$ with some $\xi \in F^*$. The induced representation $\mathcal{B}_{\chi,u}$ contains exactly one non-trivial proper invariant subspace, and the representation on it by right translation is isomorphic to the positive Weil representation with character $\psi^{u\xi}$. The representation on the factor space is isomorphic to $\mathcal{B}_{\xi,u}$.

Consequently we have the following complete list of irreducible, admissible, genuine, non-supercuspidal representations of Mp.

5.3.4 Definition.

i) If $\chi^2 \neq |\,|$ and $\chi^2 \neq |\,|^{-1}$, then $\rho_{\chi,u}$ is called a **principal series representation** . and denoted by $\tilde{\pi}_{\chi,u}$. If $m \in F^*$ is fixed and $u = -m$, then we denote it by $\tilde{\pi}_\chi$.

ii) If $\chi = | \, |^{1/2}(\cdot, \xi)$, then the representation on $\mathcal{B}_{\xi,u}$ is called a **special representation** and denoted by $\tilde{\sigma}_{\xi,u}$. For fixed $m \in F^*$ and $u = -m$ it is denoted by $\tilde{\sigma}_\xi$. The positive Weil representation on $\mathcal{B}_{\chi,u}/\mathcal{B}_{\xi,u}$ with character $\psi^{u\xi}$ is denoted by $\pi_W^{u\xi+}$.

We will see later (in Sections 5.6 and 5.8) what equivalences there are between these representations. A comment on the irritating appearance of the character ψ^{-m} will be made in Remark 5.3.6 below.

To be precise, the above theorem does not appear in this form in [Wa1]. This is because Waldspurger restricts to the case $|\chi| = | \, |^\alpha$ with $\alpha \geq 0$, which can be done in view of the following lemma ([Wa1], lemme 4).

5.3.5 Lemma. *The contragredient representation of $\mathcal{B}_{\chi,u}$ is $\mathcal{B}_{\chi^{-1}(\cdot,-1),u}$ (where $(\cdot,\cdot)$ denotes the Hilbert symbol).*

(We make the usual misuse of notation and identify representations with their representation spaces).

We give a **proof** of case iii) in the above theorem, which is not treated explicitly by Waldspurger. Let $\chi = | \, |^{-1/2}(\cdot, \xi)$ with $\xi \in F^*$. By the preceding lemma we have

$$\mathcal{B}_\chi \simeq \check{\mathcal{B}}_{\chi^{-1}(\cdot,-1)} = \check{\mathcal{B}}_{| \, |^{1/2}(\cdot,-\xi)}.$$

(we omit the u from the notation, it is fixed). It follows that the invariant subspaces of $\mathcal{B}_\chi$ correspond to the invariant quotients of $\mathcal{B}_{| \, |^{1/2}(\cdot,-\xi)}$. By part ii) of the theorem, there exists a unique proper non-trivial subspace $\mathcal{B}'_\xi$ of $\mathcal{B}_\chi$, and the following holds:

$$\mathcal{B}'_\xi \simeq \left(\mathcal{B}_{| \, |^{1/2}(\cdot,-\xi)}/\mathcal{B}_{-\xi} \right)^\vee \simeq \left(\pi_W^{-\xi u+} \right)^\vee \simeq \pi_W^{\xi u+}.$$

Now, it will be proved in Proposition 5.6.4 that there is always a non-zero intertwining map

$$\mathcal{B}_\chi \longrightarrow \mathcal{B}_{\chi^{-1}}. \tag{5.5}$$

If, for our χ, this map were injective, then

$$\pi_W^{\xi u+} \simeq \mathcal{B}'_\xi \simeq \mathcal{B}_\xi = \sigma_\xi$$

would hold, which is not the case. Hence (5.5) has the kernel $\mathcal{B}'_\xi$, and

$$\mathcal{B}_\chi/\mathcal{B}'_\xi \simeq \mathcal{B}_\xi. \qquad \square$$

Among the supercuspidal representations there are the negative (odd) Weil representations

$$\pi_W^{u-}, \qquad u \in F^*/F^{*2},$$

defined with the character ψ^u. This is proved in [Ge2] 5.5. They are very important and should be mentioned explicitly instead of just putting them in with the supercuspidals. Indeed, they turn out to be much more important then the π_w^{u+}, which do not appear as local factors in automorphic representations of Mp (Proposition 23 in [Wa1]).

We now fix an $m \in F^*$, and come up with the following complete list of irreducible, admissible, genuine representations of Mp(F).

- Those supercuspidal representations which are not equal to negative Weil representations.

- The principal series representations π_χ with $\chi^2 \neq ||^{\pm 1}$.

- The special representations σ_ξ with $\xi \in F^*/F^{*2}$.

- The positive (even) Weil representations $\pi_w^{-m\xi+}$ with $\xi \in F^*/F^{*2}$.

- The negative (odd) Weil representations $\pi_w^{-m\xi-}$ with $\xi \in F^*/F^{*2}$.

5.3.6 Remark. a) One should always keep in mind that the symbols π_χ and σ_ξ depend on $m \in F^*$, and that everything depends on the underlying character ψ. Such a dependence on the choice of a character seems to complicate matters, but is indeed an important feature of the metaplectic theory, as becomes apparent while reading [Wa1]. For the Jacobi theory, the character has the following significance. The Schrödinger-Weil representation π_{sw}^m is constructed with underlying character ψ^m. It has a certain cocycle λ, since it is projective. The metaplectic representations from above are defined with underlying character ψ^{-m}, and consequently, they have a cocycle exactly inverse to λ. Tensorizing π_{sw}^m with such a representation thus makes the two cocycles cancel, and yields a non-projective representation of the Jacobi group.

b) If ξ runs through F^*/F^{*2}, then $-m\xi$ certainly does also. In the above list we could therefore simply have written $\pi_w^{\xi\pm}$ instead of $\pi_w^{-m\xi\pm}$. But we prefer the more complicated form since it makes reference to the character ψ^{-m} and is consistent with the symbol σ_ξ, cf. Theorem 5.3.3 ii).

5.4 Induced representations

Now we define induced representations for the Jacobi group. Let χ be a character of A, and let $\mathcal{B}_{\chi,m}^J$ be the space of functions $\Phi : G^J \to \mathbb{C}$ with the following properties.

i) Φ is right invariant by some open subgroup of G^J.

ii) For every $a \in F^*$, $x, \mu, \kappa \in F$ and $g \in G^J$ the following holds:

$$\Phi\left(\begin{pmatrix} a & x \\ 0 & a^{-1} \end{pmatrix}(0, \mu, \kappa)g\right) = \chi(a)\psi^m(\kappa)|a|^{3/2}\Phi(g).$$

G^J operates on $\mathcal{B}^J_{\chi,m}$ by right translation. This is exactly the representation of G^J induced by the character

$$\begin{pmatrix} a & 0 \\ 0 & a^{-1} \end{pmatrix}(0,0,\kappa) \longmapsto \chi(a)\psi^m(\kappa)$$

of AZ. Theorem 5.4.2 below relates $\mathcal{B}^J_{\chi,m}$ to induced representations for the metaplectic group as they were introduced in the last section. We need a lemma for preparation.

5.4.1 Lemma. *Let U be a space of locally constant functions $\Phi : G^J \to \mathbb{C}$ with*

$$\Phi((0,\mu,\kappa)g) = \psi^m(\kappa)\Phi(g) \qquad \text{for all } \mu,\kappa \in F,\ g \in G^J.$$

Assume further that U is invariant under right translation, and that the representation of H on U defined by right translation (with elements of H) is equivalent to π^m_s. Then there is a locally constant function $\varphi : G \to \mathbb{C}$ such that every $\Phi \in U$ is of the form

$$\Phi(Mh) = \varphi(M)W(Mh)$$

with some W (depending on Φ) in the Whittaker model $\mathcal{W}^{m,0}_{sw}$ for π^m_{sw} with trivial character.

Proof: For any $M \in G$ the map which associates to every $\Phi \in U$ the function $h \mapsto \Phi(Mh)$ on H is an intertwining map for right translation with H on both sides. By Corollary 5.2.3, the image is either the zero space or the space $\mathcal{W}_M$ defined in this corollary. We further have the isomorphism

$$\mathcal{W}^{m,0}_{sw} \longrightarrow \mathcal{W}_M, \qquad W \longmapsto \left(h \mapsto W(Mh)\right),$$

which is also an intertwining map for the action of H. Note the image is indeed $\mathcal{W}_M$ by Corollary 5.2.3. Finally there is the isomorphism

$$\mathcal{W}^{m,0}_{sw} \xrightarrow{\sim} \mathcal{W}_1$$

given by restriction. Now we fix M and compare the map

$$U \longrightarrow \mathcal{W}_M$$

from the beginning to the composition

$$U \longrightarrow \mathcal{W}_1 \longrightarrow \mathcal{W}^{m,0}_{sw} \longrightarrow \mathcal{W}_M.$$

Everything in sight intertwines the H-action, and consequently, by Schur's lemma, these two maps differ only by a scalar, which we call $\varphi(M)$. Going through the definitions we see that for every $\Phi \in U$ and $h \in H$

$$\Phi(Mh) = \varphi(M)W(Mh)$$

with some $W \in \mathcal{W}^{m,0}_{sw}$ depending only on Φ. Given such $W \neq 0$ and $M \in G$, one sees by the construction of $\mathcal{W}^{m,0}_{sw}$, that there exists a $h \in H$ such that $W(Mh) \neq 0$. This proves finally that φ is locally constant. $\qquad\square$

5.4.2 Theorem. Let χ be a character of A and $m \in F^*$. Let $\mathcal{B}_\chi$ be the representation of Mp introduced in the last section, where the reference character is fixed to ψ^{-m}, and $\mathcal{B}_{\chi,m}^J$ the representation of G^J defined at the beginning of the present section. If $\mathcal{S}(F)$ is as usual taken as a model for π_{SW}^m, then there is a canonical isomorphism

$$\mathcal{B}_\chi \otimes \mathcal{S}(F) \longrightarrow \mathcal{B}_{\chi,m}^J, \tag{5.6}$$

$$\varphi \otimes f \longmapsto \Big(Mh \mapsto \varphi(M,1)W_f(Mh)\Big),$$

which commutes with the G^J-operation. Here W_f denotes the Whittaker function in the model $\mathcal{W}_{SW}^{m,0}$ corresponding to $f \in \mathcal{S}(F)$.

Proof: The functions in $\mathcal{B}_\chi$ transform according to

$$\varphi\left(\begin{pmatrix} a & x \\ 0 & a^{-1} \end{pmatrix}M,1\right) = \lambda\left(\begin{pmatrix} a & 0 \\ 0 & a^{-1} \end{pmatrix},M\right)\chi(a)\delta_{-m}(a)|a|\varphi(M).$$

For $W \in \mathcal{W}_{SW}^{m,0}$ we have the transformation formula (5.1). Now using property ii) from Lemma 5.3.2 we see at once that the map (5.6) is well defined. Now we show the surjectivity. By the Stone–von Neumann theorem $\mathcal{B}_{\chi,m}^J$ decomposes after restriction to H into irreducible representations U isomorphic to π_s^m. Any such U fulfills the hypotheses of the preceding lemma. So we just have to check that with the function φ appearing in this lemma the function

$$(g,\varepsilon) \longmapsto \varepsilon\varphi(g)$$

on $\widetilde{G^J}$ lies in $\mathcal{B}_\chi$. This is easily done using the fact that for every $W \in \mathcal{W}_{SW}^{m,0}$ and every $M \in G$ there exists an $h \in H$ such that $W(Mh) \neq 0$. Hence we have shown the surjectivity of our map. For the injectivity one can reason as follows. Since $\pi\big|_H$ is isotypical, there is a natural isomorphism

$$\mathrm{Hom}_H(\mathcal{S}(F),\mathcal{B}_{\chi,m}^J) \otimes \mathcal{S}(F) \xrightarrow{\sim} \mathcal{B}_{\chi,m}^J. \tag{5.7}$$

If we associate to any $\varphi \in \mathcal{B}_\chi$ the H-intertwining map

$$\mathcal{S}(F) \longrightarrow \mathcal{B}_{\chi,m}^J,$$

$$f \longmapsto \Big(Mh \mapsto \varphi(M,1)W_f(h)\Big),$$

we get an obviously injective map

$$\mathcal{B}_\chi \longrightarrow \mathrm{Hom}_H(\mathcal{S}(F),\mathcal{B}_{\chi,m}^J).$$

Tensorizing yields from this an injection

$$\mathcal{B}_\chi \otimes \mathcal{S}(F) \longrightarrow \mathrm{Hom}_H(\mathcal{S}(F),\mathcal{B}_{\chi,m}^J) \otimes \mathcal{S}(F).$$

The composition of this injection with the isomorphism (5.7) is nothing but the map in the theorem, which is therefore injective. $\qquad\square$

Remark: The occurrence of Whittaker models in the canonical isomorphism mentioned in the theorem might at first glance seem strange. But one should remember that Proposition 5.2.6 identifies $\mathcal{W}^{m,0}_{SW}$ as a representation space induced by a character of a torus, and thus this Whittaker model fits into the picture.

5.4.3 Lemma. *Let V be a smooth representation of Mp and take $\mathcal{S}(F)$ as the space for π^m_{SW}. Then the G^J-invariant subspaces of $V \otimes \mathcal{S}(F)$ are precisely those of the form $W \otimes \mathcal{S}(F)$ with W a Mp-invariant subspace of V.*

Proof: Let $U \subset V \otimes \mathcal{S}(F)$ be G^J-invariant. Define

$$W = \{v \in V : v \otimes \mathcal{S}(F) \subset U\}.$$

It is trivial that $W \otimes \mathcal{S}(F) \subset U$. But even equality holds, because of the Stone–von Neumann theorem and the interpretation of V as $\mathrm{Hom}_H(\mathcal{S}(F), V \otimes \mathcal{S}(F))$. $\qquad\square$

Theorem 5.4.2 together with the above lemma and the classification of the induced representations of Mp in Theorem 5.3.3 yields the following classification of the representations $\mathcal{B}^J_{\chi,m}$ of G^J.

5.4.4 Theorem.

i) If $\chi^2 \neq |\,|$ and $\chi^2 \neq |\,|^{-1}$, then $\mathcal{B}^J_{\chi,m}$ is irreducible.

ii) If $\chi^2 = |\,|^{\pm 1}$, then one can find $\xi \in F^*$ such that $\chi = |\,|^{\pm 1/2}(\cdot,\xi)$, where $(\,,\,)$ is the Hilbert symbol. $\mathcal{B}^J_{\chi,m}$ contains exactly one non-trivial proper invariant subspace $\mathcal{B}^J_{\pm,\xi,m}$. We have

$$\mathcal{B}^J_{\mp,\xi,m} \simeq \mathcal{B}^J_{|\,|^{\pm 1/2}(\cdot,\xi),m} \Big/ \mathcal{B}^J_{\pm,\xi,m}.$$

The representation on $\mathcal{B}^J_{|\,|^{1/2}(\cdot,\xi),m} \Big/ \mathcal{B}^J_{+,\xi,m}$ resp. on $\mathcal{B}^J_{-,\xi,m}$ is isomorphic to $\pi^m_{SW} \otimes \pi^{-m\xi+}_W$.

5.4.5 Definition.

i) If $\chi^2 \neq |\,|$ and $\chi^2 \neq |\,|^{-1}$, then the representation on $\mathcal{B}^J_{\chi,m}$ is called a **principal series representation** and denoted by $\pi_{\chi,m}$.

ii) If $\chi = |\,|^{1/2}(\cdot,\xi)$, then the representation on $\mathcal{B}^J_{+,\xi,m}$ is called a **special representation** and denoted by $\sigma_{\xi,m}$. The representation on the quotient $\mathcal{B}^J_{\chi,m}/\mathcal{B}^J_{+,\xi,m}$ is called a **positive Weil representation** and is denoted by $\sigma^+_{\xi,m}$.

5.4.6 Proposition. *The contragredient representation to* $\mathcal{B}_{\chi,m}^J$ *is* $\mathcal{B}_{\chi^{-1},-m}^J$.

Proof: By Lemma 5.3.5 the representation $\mathcal{B}_{\chi^{-1}(\cdot,-1),m}$ is contragredient to $\mathcal{B}_{\chi,m}$, and by (ii) in Lemma 5.3.2 is identical to $\mathcal{B}_{\chi^{-1},-m}$. This means that there is a non-degenerate bilinear Mp-invariant pairing

$$\langle\,,\,\rangle_0 : \mathcal{B}_{\chi,m} \times \mathcal{B}_{\chi^{-1},-m} \longrightarrow \mathbb{C}.$$

Consider further the bilinear non-degenerate pairing

$$\langle\,,\,\rangle_1 : \mathcal{S}(F) \times \mathcal{S}(F) \longrightarrow \mathbb{C},$$
$$(f\,,\,f') \longmapsto \int_F f(x)f'(x)\,dx.$$

If G^J acts on the left factor $\mathcal{S}(F)$ by π_{SW}^m and on the right by π_{SW}^{-m}, then $\langle\,,\,\rangle_1$ is G^J-invariant. Taking these two pairings together we get a third one on the tensor products

$$\mathcal{B}_{\chi,m} \otimes \mathcal{S}(F) \times \mathcal{B}_{\chi^{-1},-m} \otimes \mathcal{S}(F) \longrightarrow \mathbb{C},$$
$$(\varphi \otimes f\,,\,\varphi' \otimes f') \longmapsto \langle\varphi,\varphi'\rangle_0\langle f,f'\rangle_1.$$

This is easily seen to be G^J-invariant and non-degenerate, if G^J acts on the first and the second factor $\mathcal{S}(F)$ as before. The assertion follows now from Theorem 5.4.2. □

5.5 Supercuspidal representations

In Section 5.3 we already defined the notion of supercuspidal representation for the metaplectic group. Now we make an analogous definition for the Jacobi group. So let π be an admissible representation of $G^J(F)$ on a vector space V. Then we call π **supercuspidal** (relative to N^J) if for every $v \in V$ there exists an open compact subgroup N of N^J such that

$$\int_N \pi(n)v\,dn = 0.$$

We look at representations π of G^J of the form

$$\pi = \tilde{\pi} \otimes \pi_{SW}^m,$$

where $\tilde{\pi}$ is an admissible representation of Mp, considered as a projectice representation of G with cocycle λ.

5.5.1 Proposition. *For π as above we have*

$$\pi \text{ supercuspidal} \quad\Longleftrightarrow\quad \tilde{\pi} \text{ supercuspidal}.$$

Proof: Assume $\tilde{\pi}$ supercuspidal. For given $v \in V$ there exists an integer r such that

$$\int_{\omega^r \mathcal{O}} \pi \begin{pmatrix} 1 & x \\ 0 & 1 \end{pmatrix} v \, dx = 0.$$

If there is further given an $f \in \mathcal{S}(F)$, our standard model for π_{SW}^m, we can find an integer s such that for every $y \in F$ the integral

$$\int_{\omega^s \mathcal{O}} \psi\left(2my\left(\mu + \frac{1}{2}xy\right)\right) f(y) \, d\mu$$

is independent from $x \in \omega^r \mathcal{O}$, because f has compact support. We conclude that

$$\int_{\omega^s \mathcal{O}} \pi_{SW}^m \left(\begin{pmatrix} 1 & x \\ 0 & 1 \end{pmatrix}(0, \mu, 0)\right) f \, d\mu \qquad \text{does not depend on } x \in \omega^r \mathcal{O}.$$

Therefore

$$\int_{\omega^r \mathcal{O}} \int_{\omega^s \mathcal{O}} (\tilde{\pi} \otimes \pi_{SW}^m)\left(\begin{pmatrix} 1 & x \\ 0 & 1 \end{pmatrix}(0, \mu, 0)\right)(v \otimes f) \, dx \, d\mu$$

$$= \int_{\omega^r \mathcal{O}} \int_{\omega^s \mathcal{O}} \left(\tilde{\pi}\begin{pmatrix} 1 & x \\ 0 & 1 \end{pmatrix} v\right) \otimes \left(\pi_{SW}^m\left(\begin{pmatrix} 1 & x \\ 0 & 1 \end{pmatrix}(0, \mu, 0)\right) f\right) \, dx \, d\mu$$

$$= \int_{\omega^r \mathcal{O}} \int_{\omega^s \mathcal{O}} \left(\tilde{\pi}\begin{pmatrix} 1 & x \\ 0 & 1 \end{pmatrix} v\right) \otimes \left(\pi_{SW}^m(0, \mu, 0) f\right) \, dx \, d\mu$$

$$= \int_{\omega^r \mathcal{O}} \tilde{\pi}\begin{pmatrix} 1 & x \\ 0 & 1 \end{pmatrix} v \, dx \otimes \int_{\omega^s \mathcal{O}} \pi_{SW}^m(0, \mu, 0) f \, d\mu = 0.$$

This shows that $\pi = \tilde{\pi} \otimes \pi_{SW}^m$ is supercuspidal. Assume conversely this is true. Given $v \in V$, we set $f = 1_{\mathcal{O}}$ and find integers r, s such that

$$\int_{\omega^r \mathcal{O}} \int_{\omega^s \mathcal{O}} (\tilde{\pi} \otimes \pi_{SW}^m)\left(\begin{pmatrix} 1 & x \\ 0 & 1 \end{pmatrix}(0, \mu, 0)\right)(v \otimes f) \, dx \, d\mu = 0.$$

This equation remains true if we shift s far into the negatives. Then the above argument shows again that the right side of the tensor is independent from $x \in \omega^r \mathcal{O}$, and we arrive at

$$\int_{\omega^r \mathcal{O}} \tilde{\pi}\begin{pmatrix} 1 & x \\ 0 & 1 \end{pmatrix} v \, dx \otimes \int_{\omega^s \mathcal{O}} \pi_{SW}^m(0, \mu, 0) f \, d\mu = 0.$$

But the Schwartz function

$$\int\limits_{\omega^s \mathcal{O}} \pi_{SW}^m(0, \mu, 0) f \, d\mu$$

does not vanish at 0, hence the first integral equals zero. This proves that $\tilde{\pi}$ is supercuspidal. $\qquad\square$

Just like in the reductive theory, we have the following **subrepresentation theorem**.

5.5.2 Proposition. *If an irreducible, admissible representation of G^J (with central character ψ^{-m}) is not supercuspidal, then it is a principal series, special, or positive Weil representation, i.e., it is a subquotient of some $\mathcal{B}_{\chi,m}^J$.*

Proof: Given π an irreducible, admissible representation of G^J, we can write it as $\pi = \tilde{\pi} \otimes \pi_{SW}^m$ with an irreducible, admissible representation $\tilde{\pi}$ of Mp. If π is not supercuspidal, then $\tilde{\pi}$ is also not. But for the metaplectic group the subrepresentation theorem holds, cf. [Ge2], Section 5.1, which means that $\tilde{\pi}$ is a subrepresentation or a quotient of some $\mathcal{B}_\chi$. In view of Theorem 5.4.2, the assertion for the Jacobi group follows. $\qquad\square$

From Proposition 5.5.1 and the remarks made after Lemma 5.3.5, we see that there exist some supercuspidal representations of G^J which deserve special attention, namely the representations

$$\sigma_{\xi,m}^- := \pi_{SW}^m \otimes \pi_W^{-m\xi-}$$

coming from negative Weil representations of the metaplectic group. We call them also **negative Weil representations** of G^J. They will appear in our final classification of the irreducible, admissible representations of $G^J(F)$ in Section 5.8.

5.6 Intertwining operators

The results of the last two sections give an overview over the irreducible, admissible representations of G^J. For a complete classification, it remains to describe the equivalences between them. This section makes a step in this direction by introducing intertwining operators between representations which one would expect to be equivalent. After that we will show that there are no other equivalences except these obvious ones, by utilizing Whittaker and Kirillov models.

First we deal with the metaplectic group and its induced representations $\mathcal{B}_\chi$, as they were defined in Section 5.3. From the general theory for reductive groups one expects that χ and χ^{-1} yield the same representation of Mp. We are proving this fact by writing down an intertwining operator very similar as in the case of SL(2). Namely, for $\varphi \in \mathcal{B}_\chi$ we define

$$(\tilde{I}\varphi)(g) = \int_F \varphi\left((w,1)\begin{pmatrix} 1 & x \\ 0 & 1 \end{pmatrix} g\right) dx, \qquad g \in \mathrm{Mp}. \tag{5.8}$$

5.6.1 Proposition. *Let χ be a character of F^* and $\sigma \in \mathbb{R}$ be defined by $|\chi(a)| = |a|^\sigma$. If $\sigma > 0$, then the integral (5.8) converges absolutely for every $g \in G$, the function $\tilde{I}\varphi$ lies in $\mathcal{B}_{\chi^{-1}}$, and the association $\varphi \mapsto \tilde{I}\varphi$ is a nonzero intertwining map $\mathcal{B}_\chi \to \mathcal{B}_{\chi^{-1}}$.*

Proof: For the question of convergence we note that

$$
\left|\varphi\left((w,1)\begin{pmatrix} 1 & x \\ 0 & 1 \end{pmatrix} g\right)\right| = \left|\varphi\left(\left(w\begin{pmatrix} 1 & x \\ 0 & 1 \end{pmatrix}, 1\right) g\right)\right|
$$

$$
= \left|\varphi\left(\left(\begin{pmatrix} -x^{-1} & 1 \\ 0 & -x \end{pmatrix}\begin{pmatrix} 1 & 0 \\ x^{-1} & 1 \end{pmatrix}, 1\right) g\right)\right|
$$

$$
= |x|^{-1}|x|^{-\sigma} \left|\varphi\left(\left(\begin{pmatrix} 1 & 0 \\ x^{-1} & 1 \end{pmatrix}, 1\right) g\right)\right|.
$$

If the absolute value of x tends to infinity, then the elements

$$
\left(\begin{pmatrix} 1 & 0 \\ x^{-1} & 1 \end{pmatrix}, 1\right)
$$

converge to the identity in Mp (if this should not be clear because of the strange topology in Mp, conjugate by $(w,1)$ and use the fact that N is contained in Mp). Therefore we have for x large enough

$$
\left|\varphi\left((w,1)\begin{pmatrix} 1 & x \\ 0 & 1 \end{pmatrix} g\right)\right| = |x|^{-1}|x|^{-\sigma} |\varphi(g)|,
$$

so that the absolute convergence of (5.8) is equivalent with the convergence of

$$
\int_{|x|>c} |x|^{-\sigma-1} dx \qquad (c > 0).
$$

But it is a standard p-adic computation to show that the latter integral is finite exactly for $\sigma > 0$.

We have to show that indeed $\tilde{I}\varphi \in \mathcal{B}_{\chi^{-1}}$. It is clear that

$$
\tilde{I}\varphi\left(\begin{pmatrix} 1 & x \\ 0 & 1 \end{pmatrix} g\right) = \tilde{I}\varphi(g) \qquad \text{for all } x \in F, \ g \in \mathrm{Mp},
$$

so it remains to prove that for all $a \in F^*$, $\varepsilon \in \{\pm 1\}$ and $g \in \mathrm{Mp}$

$$
\varphi\left(\left(\begin{pmatrix} a & 0 \\ 0 & a^{-1} \end{pmatrix}, \varepsilon\right) g\right) = \varepsilon \delta_m(a)|a|\chi^{-1}(a)\varphi(g).
$$

But this is straightforward and may therefore be left to the reader.

It is clear now by the definition that $\tilde{I}$ intertwines $\mathcal{B}_\chi$ with $\mathcal{B}_{\chi^{-1}}$. Finally it has to be shown that $\tilde{I}$ is not the zero operator. For this purpose we utilize the Bruhat decomposition in $SL(2, F)$ which says $SL(2, F) = NA \cup NAwN$ (disjoint). A function φ on Mp can therefore be defined by

$$\varphi\left(\begin{pmatrix} a & b \\ 0 & a^{-1} \end{pmatrix} w \begin{pmatrix} 1 & x \\ 0 & 1 \end{pmatrix}, \varepsilon\right) = \varepsilon \delta_m(a) \chi(a)|a| \qquad \text{if } a \in F^*, \ b \in F, \ x \in \mathcal{O},$$

and $\varphi(g) = 0$ if g is not of the indicated form. Then it is obvious that φ lies in $\mathcal{B}_\chi$, and that $\tilde{I}\varphi$ does not vanish at $g = 1$. So $\tilde{I}$ is not zero. $\qquad\square$

Let us still assume that $|\chi| = |\ |^\sigma$ with $\sigma > 0$. Then the intertwining operator $\tilde{I} : \mathcal{B}_\chi \to \mathcal{B}_{\chi^{-1}}$ gives via the isomorphism

$$\mathcal{B}_\chi \otimes \mathcal{S}(F) \xrightarrow{\ \sim\ } \mathcal{B}^J_{\chi,m}$$

rise to an operator

$$I : \mathcal{B}^J_{\chi,m} \longrightarrow \mathcal{B}^J_{\chi^{-1},m},$$

which we shall now compute. For this purpose let $\Phi \in \mathcal{B}^J_{\chi,m}$ be of the form

$$\Phi(Mh) = \varphi(M, 1) W_f(Mh) \qquad \text{for all } M \in G, \ h \in H,$$

where $\varphi \in \mathcal{B}_\chi$, $f \in \mathcal{S}(F)$, and $W_f \in \mathcal{W}^{m,0}_{sw}$ the Whittaker function corresponding to f. Then

$$(I\Phi)(Mh) = (\tilde{I}\varphi)(M, 1) W_f(Mh)$$

$$= \int_F \varphi\left((w, 1)\begin{pmatrix} 1 & x \\ 0 & 1 \end{pmatrix}(M, 1)\right) W_f(Mh)\, dx$$

$$= \lambda\left(w\begin{pmatrix} 1 & x \\ 0 & 1 \end{pmatrix}, M\right) \lambda\left(\begin{pmatrix} 1 & x \\ 0 & 1 \end{pmatrix}, M\right)$$

$$\cdot \int_F \varphi\left(w\begin{pmatrix} 1 & x \\ 0 & 1 \end{pmatrix}M, 1\right) W_f\left(\begin{pmatrix} 1 & x \\ 0 & 1 \end{pmatrix}Mh\right) dx$$

$$= \lambda\left(w\begin{pmatrix} 1 & x \\ 0 & 1 \end{pmatrix}, M\right)$$

$$\cdot \int_F \varphi\left(w\begin{pmatrix} 1 & x \\ 0 & 1 \end{pmatrix}M, 1\right) W_f\left(w^{-1}w\begin{pmatrix} 1 & x \\ 0 & 1 \end{pmatrix}Mh\right) dx. \qquad (5.9)$$

Now we utilize the following lemma, which for later use is stated in greater generality than presently required.

5.6.2 Lemma. Let $\mathcal{W}_{sw}^{m,r}$ be the $\psi^{\nu,r}$-Whittaker model with parameters $\frac{r^2}{4m}$ and $r \in F$ for the Schrödinger-Weil representation (cf. Theorem 5.2.4). Then for every $M \in G$, $h \in H$ and $W \in \mathcal{W}_{sw}^{m,r}$ the following holds:

$$W(w^{-1}Mh) = \lambda(w^{-1}, M)\gamma_m(-1) \int_F W((\mu, 0, 0)Mh)\psi(-r\mu)\, d\mu.$$

Proof: Let $W = W_f$ with $f \in \mathcal{S}(F)$. Then one computes, using explicit formulas for λ and the Schrödinger-Weil representation,

$$
\begin{aligned}
W(w^{-1}Mh) &= \left(\pi_{sw}^m\left(\begin{pmatrix} -1 & 0 \\ 0 & -1 \end{pmatrix}wMh\right)f\right)\left(\frac{r}{2m}\right) \\
&= \lambda(-1, wM)\left(\pi_{sw}^m(-1)\pi_{sw}^m(wMh)f\right)\left(\frac{r}{2m}\right) \\
&= \lambda(-1, wM)\lambda(w, M)\delta_m(-1)\left(\pi_{sw}^m(w)\pi_{sw}^m(Mh)f\right)\left(\frac{-r}{2m}\right) \\
&= \lambda(w^{-1}, M)\lambda(-1, w)\delta_m(-1)\gamma_m(1)\int_F \left(\pi_{sw}^m(Mh)f\right)(\mu)\psi(-r\mu)\, d\mu \\
&= \lambda(w^{-1}, M)(-1, -1)\delta_m(-1)\gamma_m(1) \\
&\qquad\qquad \int_F \left(\pi_{sw}^m((\mu, 0, 0)Mh)f\right)(0)\psi(-r\mu)\, d\mu \\
&= \lambda(w^{-1}, M)\gamma_m(-1)\int_F W((\mu, 0, 0)Mh)\psi(-r\mu)\, d\mu. \qquad \square
\end{aligned}
$$

Now, using this lemma for $r = 0$ and $w\begin{pmatrix} 1 & x \\ 0 & 1 \end{pmatrix}M$ instead of M, we can go on with the calculation (5.9).

$$
\begin{aligned}
(I\Phi)(Mh) &= (\tilde{I}\varphi)(M, 1)W_f(Mh) \\
&= \gamma_m(-1)\lambda\left(w, \begin{pmatrix} 1 & x \\ 0 & 1 \end{pmatrix}M\right)\lambda\left(w^{-1}, w\begin{pmatrix} 1 & x \\ 0 & 1 \end{pmatrix}M\right) \\
&\qquad \int_F \int_F \varphi\left(w\begin{pmatrix} 1 & x \\ 0 & 1 \end{pmatrix}M, 1\right)W_f\left((\mu, 0, 0)w\begin{pmatrix} 1 & x \\ 0 & 1 \end{pmatrix}Mh\right)d\mu\, dx \\
&= \gamma_m(-1)\lambda(w^{-1}, w)\lambda\left(1, w\begin{pmatrix} 1 & x \\ 0 & 1 \end{pmatrix}M\right) \\
&\qquad \int_F \int_F \varphi\left(w\begin{pmatrix} 1 & x \\ 0 & 1 \end{pmatrix}M, 1\right)W_f\left(w\begin{pmatrix} 1 & x \\ 0 & 1 \end{pmatrix}(0, \mu, 0)Mh\right)d\mu\, dx \\
&= \gamma_m(-1)\int_F \int_F \Phi\left(w\begin{pmatrix} 1 & x \\ 0 & 1 \end{pmatrix}(0, \mu, 0)Mh\right)d\mu\, dx.
\end{aligned}
$$

Up to the factor $\gamma_m(-1)$, this is the exact analogue to the definition (5.8), because the double integral may be interpreted as integration over N^J. But it is not hard to see that this time the convergence is not absolute, hence the order of integration must not be changed. In view of Proposition 5.6.1 we have proved the following.

5.6.3 Proposition. *Let* $|\chi| = ||^\sigma$ *with* $\sigma > 0$. *For any* $\Phi \in \mathcal{B}^J_{\chi,m}$ *and* $g \in G^J$ *the integral*

$$(I\Phi)(g) = \gamma_m(-1) \int\limits_F \int\limits_F \Phi\left(w\begin{pmatrix} 1 & x \\ 0 & 1 \end{pmatrix}(0,\mu,0)g\right) d\mu\, dx$$

is convergent, but not absolutely convergent. The map $\Phi \mapsto I\Phi$ *is a nonzero intertwining map* $\mathcal{B}^J_{\chi,m} \to \mathcal{B}^J_{\chi^{-1},m}$, *which coincides with the operator* $\tilde{I} \otimes \mathrm{id}$ *via the isomorphism* $\mathcal{B}_\chi \otimes \mathcal{S}(F) \xrightarrow{\sim} \mathcal{B}^J_{\chi,m}$.

If $\mathcal{B}^J_{\chi,m}$ and $\mathcal{B}^J_{\chi^{-1},m}$ are both irreducible, and if $|\chi| = ||^\sigma$ with $\sigma > 0$, then this proposition implies that $\mathcal{B}^J_{\chi,m} \simeq \mathcal{B}^J_{\chi^{-1},m}$. The same is true if $\sigma < 0$ by interchanging the roles of χ and χ^{-1}. It remains to treat the case $\sigma = 0$, which indeed turns out to be the most interesting one, because these representations constitute the unitary principal series.

We can write our character χ (in a non-unique way) as

$$\chi(a) = |a|^s \chi_0(a) \qquad (a \in F^*),$$

where χ_0 is a unitary character of F^* and $s \in \mathbb{C}$. The above σ is nothing but $\mathrm{Re}(s)$. We can get into the region $\sigma = 0$ by viewing the intertwining integrals as functions of s in the domain $\mathrm{Re}(s) > 0$, and then continue analytically. This will be done here on both sides of the isomorphism $\mathcal{B}_\chi \otimes \mathcal{S}(F) \simeq \mathcal{B}^J_{\chi,m}$, i.e., for the metaplectic group as well as for the Jacobi group.

The metaplectic case is treated first. The Iwasawa decomposition $G = BK$ $(K = \mathrm{SL}(2,\mathcal{O}))$ implies that any $\varphi \in \mathcal{B}_\chi$ is determined by its restriction to $\tilde{K} = (K, \{\pm 1\}) \subset \mathrm{Mp}$. This restriction satisfies

$$\varphi\left(\left(\begin{pmatrix} a & x \\ 0 & a^{-1} \end{pmatrix}, \varepsilon\right) k\right) = \varepsilon \delta_m(a) \chi_0(a) \varphi(k)$$

for $a \in \mathcal{O}^*$, $x \in \mathcal{O}$, $\varepsilon \in \{\pm 1\}$, $k \in \tilde{K}$. Define V to be the space of all locally constant functions $\tilde{K} \to \mathbb{C}$ having this property. Now if $\varphi \in V$ is fixed, then it is obvious that for every $s \in \mathbb{C}$ there exists a unique extension of φ to a function $\varphi_s \in \mathcal{B}_{\chi_0||^s}$. As in [Bu] p. 350 we refer to the map $s \mapsto \varphi_s$ as a *flat section*.

5.6.4 Proposition. Let $s \mapsto \varphi_s$ be the flat section corresponding to a fixed $\varphi \in V$. For fixed $g \in \mathrm{Mp}$ the integral

$$(\tilde{I}\varphi_s)(g) = \int_F \varphi_s\left((w,1)\begin{pmatrix} 1 & x \\ 0 & 1 \end{pmatrix} g\right) dx$$

defines a holomorphic function on the domain $\mathrm{Re}(s) > 0$. This function has analytic continuation to all s except where $\chi = \chi_0 |\,|^s = 1$, and defines a nonzero intertwining operator

$$\mathcal{B}_{\chi_0 |\,|^s} \longrightarrow \mathcal{B}_{\chi_0^{-1} |\,|^{-s}}.$$

Proof: Because of

$$(\tilde{I}\varphi_s)\left(\left(\begin{pmatrix} a & x \\ 0 & a^{-1} \end{pmatrix}, \varepsilon\right) g\right) = \varepsilon |a|^{1-s}\chi_0(a)^{-1}\delta_m(a)\varphi_s(g),$$

we may assume that $g \in \tilde{K}$. Similar to the proof of Proposition 5.6.1 we find a positive integer $N \in \mathbb{N}$ such that

$$\varphi\left(\left(\begin{pmatrix} 1 & 0 \\ x^{-1} & 1 \end{pmatrix}, 1\right) g\right) = \varphi(g) \qquad \text{if } |x| > q^N.$$

Then we split the integral $(\tilde{I}\varphi_s)(g)$ into $\int_{|x| \leq q^N}$ and $\int_{|x| > q^N}$. The first integration being over a compact set, analytic continuation is no problem. The second integral equals

$$\varphi(g) \int_{|x| \geq q^{N+1}} |x|^{-s-1}\chi_0^{-1}(x)\, dx = \varphi(g) \sum_{n=N+1}^{\infty} |\omega^{-n}|^{-s}\chi_0(\omega^n) \int_{\mathcal{O}^*} \chi_0^{-1}(x)\frac{dx}{|x|}.$$

If χ_0 is not unramified, i.e. is not trivial on $\mathcal{O}^*$, then all the integrals over $\mathcal{O}^*$ vanish, hence there is nothing to prove. Otherwise we arrive at

$$\varphi(g) \sum_{n=N+1}^{\infty} q^{-ns}\chi_0(\omega)^n = \varphi(g)(q^{-s}\chi_0(\omega))^{N+1} \sum_{n=0}^{\infty}(q^{-s}\chi_0(\omega))^n,$$

the multiplicative measure being suitably normalized. The last sum equals $(1 - q^{-s}\chi_0(\omega))^{-1}$ provided $q^{-s}\chi_0(\omega) \neq 1$. This condition is equivalent to $\chi \neq 1$, and the assertions about holomorphy and analytic continuation are proved. The fact that the function $g \mapsto (\tilde{I}\varphi_s)(g)$ lies in $\mathcal{B}_{\chi_0^{-1} |\,|^{-s}}$, that the analytically continued integral defines an intertwining operator, and that this operator is nonzero, now follow in a straightforward way from the identity theorem. $\qquad\square$

The results obtained in this proposition will now be taken over to the Jacobi group. Every $\Phi \in \mathcal{B}_{\chi,m}^J$ is determined by its restriction to KH, this restriction satisfying

$$\Phi\left(\begin{pmatrix} a & x \\ 0 & a^{-1} \end{pmatrix}(0, \mu, \kappa)g\right) = \chi_0(a)\psi^m(\kappa)\Phi(g)$$

for $a \in \mathcal{O}^*$, $x, \mu, \kappa \in F$, $g \in KH$. If V^J is the space of all such functions on KH, then for every $s \in \mathbb{C}$ any $\Phi \in V^J$ can be extended uniquely to a function $\Phi_s \in \mathcal{B}_{\chi_0||^s,m}$ by the rule

$$\Phi_s\left(\begin{pmatrix} a & x \\ 0 & a^{-1} \end{pmatrix} kh\right) = |a|^{3/2}\chi(a)\Phi(kh) \quad \text{for } a \in F^*, \ x \in F, \ k \in K, \ h \in H.$$

The map $s \mapsto \Phi_s$ is again called the *flat section* belonging to $\Phi \in V^J$.

5.6.5 Proposition. *Let $\Phi \in V^J$ and $s \mapsto \Phi_s$ the corresponding flat section. For fixed $g \in G^J$ the integral*

$$(I\Phi_s)(g) = \int\limits_{F}\int\limits_{F} \Phi_s\left(x\begin{pmatrix} 1 & x \\ 0 & 1 \end{pmatrix}(0, \mu, 0)g\right) d\mu\, dx$$

defines a holomorphic function in $\mathrm{Re}(s) > 0$. This function can be analytically continued to all s where $\chi = \chi_0||^s \neq 1$, and after that the operator I defines a nonzero intertwining operator

$$\mathcal{B}^J_{\chi_0||^s,m} \longrightarrow \mathcal{B}^J_{\chi_0^{-1}||^{-s},m}.$$

Proof: This is an easy consequence of Propositions 5.6.3 and 5.6.4. □

5.6.6 Corollary. *The principal series representations for the characters χ and χ^{-1} are equivalent:*

$$\pi_{\chi,m} \simeq \pi_{\chi^{-1},m}.$$

5.7 Whittaker models

As usual, let $\tilde{\pi}$ be an irreducible smooth representation of Mp and

$$\pi = \tilde{\pi} \otimes \pi^m_{SW}$$

be the corresponding irreducible smooth representation of G^J with central character ψ^m on a space V. We would like to realize π as a space of functions $W : G^J \to \mathbb{C}$ which transform according to

$$W\left(\begin{pmatrix} 1 & x \\ 0 & 1 \end{pmatrix}(0, \mu, 0)g\right) = \psi^n(x)\psi^r(\mu)W(g) \quad \text{for all } x, \mu \in F, \ g \in G^J,$$

where $n, r \in F$ are parameters. If such a space exists such that right translation on it defines a representation equivalent to π, then π is said to have a $\psi^{n,r}$-**Whittaker model** (cf. the corresponding notions in the real case, Section 3.6). This model will then be denoted $\mathcal{W}^{n,r}_\pi$. It is said to be unique if there is only one such space in the space of all locally constant functions on G^J. If

$v \mapsto W_v$ intertwines V with $\mathcal{W}^{n,r}_\pi$, then the corresponding **Whittaker functional** $l : V \to \mathbb{C}$, $v \mapsto W_v(1)$, has the property

$$l\left(\pi\left(\begin{pmatrix} 1 & x \\ 0 & 1 \end{pmatrix}(0,\mu,0)\right)v\right) = \psi^n(x)\psi^r(\mu)l(v) \qquad \text{for all } x, \mu \in F, \ v \in V.$$

Conversely, any such functional on V yields a $\psi^{n,r}$- Whittaker model via $W_v(g) = l(\pi(g)v)$. The existence and uniqueness of $\mathcal{W}^{n,r}_\pi$ is equivalent to the property that the space of $\psi^{n,r}$-Whittaker functionals be one-dimensional.

Very similar notions exist for representations of Mp. Here one requires the transformation property for the subgroup N of Mp. It should be clear what a ψ^ν-Whittaker model or Whittaker functional for $\tilde{\pi}$ is without stating all the details.

Whittaker models for π^m_{SW} were discussed in Section 5.2. The unique $\psi^{\nu,r}$-Whittaker functional ($\nu = r^2/4m$) for π^m_{SW} coincides with the unique ψ^r-Whittaker functional for π^m_S and is given by

$$l^{m,r}_{SW}(f) = f\left(\frac{r}{2m}\right) \qquad \text{for all } f \in \mathcal{S}(F).$$

5.7.1 Proposition. *Let $\tilde{\pi} : \mathrm{Mp} \to \mathrm{GL}(V)$ be a smooth representation, and let $\mathcal{S}(F)$ be the standard space for π^m_{SW}. The following are equivalent:*

i) *There exists a $\psi^{n,r}$-Whittaker functional l on $V \otimes \mathcal{S}(F)$.*

ii) *There exists a ψ^ν-Whittaker functional $\tilde{l}$ on V, where $\nu = n - \frac{r^2}{4m}$.*

In case of existence we have

$$l(v \otimes f) = \tilde{l}(v)l^{m,r}_{SW}(f) \qquad \text{for } v \in V, \ f \in \mathcal{S}(F).$$

So the space of $\psi^{n,r}$-Whittaker functionals for $\pi = \tilde{\pi} \otimes \pi^m_{SW}$ is isomorphic to the space of ψ^ν-Whittaker functionals for $\tilde{\pi}$, where $\nu = n - \frac{r^2}{4m}$. In particular, if $\tilde{\pi}$ is irreducible, then it has a unique ψ^ν-Whittaker model if and only if π has a unique $\psi^{n,r}$-Whittaker model.

Proof: If $\tilde{l} : V \to \mathbb{C}$ is a ψ^ν-Whittaker functional, then it is very easy to check that

$$\begin{aligned} l : V \otimes \mathcal{S}(F) &\longrightarrow \mathbb{C}, \\ v \otimes f &\longmapsto \tilde{l}(v)l^{m,r}_{SW}(f), \end{aligned}$$

defines a $\psi^{n,r}$-Whittaker functional, with $n = \nu + \frac{r^2}{4m}$. Conversely, if a $\psi^{n,r}$-Whittaker functional $l : V \otimes \mathcal{S}(F) \to \mathbb{C}$ is given, then it is obvious that for fixed $v \in V$ the linear map

$$\begin{aligned} \mathcal{S}(F) &\longrightarrow \mathbb{C}, \\ f &\longmapsto l(v \otimes f) \end{aligned}$$

defines a ψ^r-Whittaker functional for π_s^m. By Theorem 5.2.1, this functional differs from $l_{sW}^{m,r}(f)$ only by a constant $\tilde{l}(v)$ depending on v, i.e.

$$l(v \otimes f) = \tilde{l}(v) l_{sW}^{m,r}(f).$$

Now it is easy to check that $v \mapsto \tilde{l}(v)$ defines a ψ^ν-Whittaker functional for $\tilde{\pi}$, with $\nu = n - \frac{r^2}{4m}$. $\qquad\qquad\square$

By this proposition, the existence and uniqueness question for the Whittaker models for representations of G^J is completely reduced to the metaplectic case. Concerning the induced representations $\mathcal{B}_\chi$ of Mp (cf. Section 5.3) there are the following complete results of Waldspurger in [Wal], where the reference character is as before fixed to ψ^{-m}. For questions of convergence we have to ourselves to the case $|\chi| = ||^\alpha$ with $\alpha \geq 0$. This is enough in view of Proposition 5.6.4. On the space $\mathcal{B}_\chi$ consider the functional $\tilde{l}^\nu$, $\nu \in F^*$, given by

$$\tilde{l}^\nu(\varphi) = \int_F \varphi\left(w\begin{pmatrix} 1 & x \\ 0 & 1 \end{pmatrix}, 1\right) \psi^{-\nu}(x)\,dx \qquad\qquad (\varphi \in \mathcal{B}_\chi).$$

It is almost obvious that this is a ψ^ν-Whittaker functional on $\mathcal{B}_\chi$ provided it is non-zero. From [Wal], Prop. 3, p. 14, one can deduce the following:

- If $\chi^2 \neq ||$ (i.e. the principal series case), then $\tilde{l}^\nu$ is non-trivial for every $\nu \in F^*$.

- If $\chi = ||^{1/2}(\cdot, \xi)$, then the restriction of $\tilde{l}^\nu$ to $\mathcal{B}_\xi$ is non-trivial exactly for

$$\nu F^{*2} \neq -m\xi F^{*2}.$$

- If $\chi = ||^{1/2}(\cdot, \xi)$, and $\nu F^{*2} = -m\xi F^{*2}$, then the resulting functional on $\mathcal{B}_\chi/\mathcal{B}_\xi$ is non-trivial.

In any case, $\tilde{l}^\nu$ is unique up to scalars. The above proposition states that $l := \tilde{l}^\nu \otimes l_{sW}^{m,r}$ defines a Whittaker functional on $\mathcal{B}_\chi \otimes \mathcal{S}(F)$. This will be taken over now to $\mathcal{B}_{\chi,m}^J$ (cf. Theorem 5.4.2). The calculation is only a slight generalization of (5.9), so we just state the result:

$$l(\Phi) = \int_F \int_F \Phi\left(w\begin{pmatrix} 1 & x \\ 0 & 1 \end{pmatrix}(0, \mu, 0)\right) \psi^{-n}(x)\psi^{-r}(\mu)\,d\mu\,dx \qquad (5.10)$$

for $\Phi \in \mathcal{B}_{\chi,m}^J$ (up to an irrelevant constant). This is exactly what one would expect. But just like in our discussion of intertwining integrals, the convergence of (5.10) is not absolute, so that the order of integration has to be observed. Waldspurger's results, the above proposition, Theorem 5.4.2 and Corollary 5.2.5 add up to the following result.

5.7.2 Theorem. Let $n, r \in F$ and $N = 4mn - r^2$.

  i) *The principal series representations $\pi_{\chi,m}$ have a $\psi^{n,r}$-Whittaker model for all $n, r \in F$ with $N \neq 0$.*

  ii) *The special representation $\sigma_{\xi,m}$ has a $\psi^{n,r}$-Whittaker model if $N \neq 0$ and $-NF^{*2} \neq \xi F^{*2}$.*

  iii) *The Weil representations $\sigma_{\xi,m}^{\pm}$ have a $\psi^{n,r}$-Whittaker model if $N \neq 0$ and $-NF^{*2} = \xi F^{*2}$.*

In each case, if $W_{\pi}^{n,r}$ exists, it is unique. A non-trivial Whittaker functional on the standard space of the induced representation is given (resp. induced) by (5.10), provided $|\chi| = |\,|^{\alpha}$ with $\alpha \geq 0$.

Let $\tilde{\pi}$ be one of the induced representations of Mp (resp. an irreducible subquotient). Assume the Whittaker model $\mathcal{W}_{\tilde{\pi}}^{\nu}$ exists for some $\nu \in F$. It turns out to be important to consider the space of functions

$$\mathcal{K}(\tilde{\pi}) := \left\{ x \longmapsto W\left(\begin{pmatrix} a & 0 \\ 0 & a^{-1} \end{pmatrix}, 1 \right) : \ W \in \mathcal{W}_{\tilde{\pi}}^{\nu} \right\}.$$

This space of functions is called the **Kirillov model** for $\tilde{\pi}$, although it is not a representation space for $\tilde{\pi}$, due to the fact that restriction of Whittaker functions to the diagonal in general is not injective. In [Wa1], Prop. 4, Waldspurger gives the following description of $\mathcal{K}(\tilde{\pi})$ in terms of the even Schwartz space

$$\mathcal{S}(F)^{+} = \{ f \in \mathcal{S}(F) : \ \forall\, x \in F \ \ f(x) = f(-x) \}.$$

- If $\tilde{\pi} = \tilde{\pi}_{\chi,-m}$ is a principal series representation with $\chi^2 \neq 1$, then

$$\mathcal{K}(\tilde{\pi}) = \Big\{ a \mapsto \delta_{-m}(a)|a|(\chi(a)f_1(a) + \chi^{-1}(a)f_2(a)) :$$
$$f_1, f_2 \in \mathcal{S}(F)^{+} \Big\}.$$

- If $\tilde{\pi} = \tilde{\pi}_{\chi,-m}$ is a principal series representation with $\chi^2 = 1$, then

$$\mathcal{K}(\tilde{\pi}) = \Big\{ a \mapsto \delta_{-m}(a)|a|(\chi(a)f_1(a) + \chi^{-1}(a)v_F(a)f_2(a)) :$$
$$f_1, f_2 \in \mathcal{S}(F)^{+} \Big\}.$$

- If $\tilde{\pi} = \tilde{\sigma}_{\xi,-m}$ is a special representation, then

$$\mathcal{K}(\tilde{\pi}) = \Big\{ a \mapsto \delta_{-m}(a)|a|\chi(a)f(a) : \ f \in \mathcal{S}(F)^{+} \Big\}.$$

- If $\tilde{\pi} = \pi_W^{-m\xi+}$ is a positive Weil representation, then

$$\mathcal{K}(\tilde{\pi}) = \Big\{ a \mapsto \delta_{-m}(a)|a|\chi^{-1}(a)f(a) : \ f \in \mathcal{S}(F)^{+} \Big\}.$$

If π is a representation of G^J with Whittaker model $\mathcal{W}_\pi^{n,r}$, then it seems reasonable to define the Kirillov model for π as

$$\mathcal{K}(\pi) := \left\{ F^* \times F \ni (a, \lambda) \longmapsto W\left(\begin{pmatrix} a & 0 \\ 0 & a^{-1} \end{pmatrix}(\lambda, 0, 0) \right) : W \in \mathcal{W}_\pi^{n,r} \right\}.$$

We know that Whittaker functions for $\pi = \tilde{\pi} \otimes \pi_{SW}^m$ are combined from Whittaker functions for $\tilde{\pi}$ and those for π_{SW}^m. Now for any $W = W_f \in \mathcal{W}_{SW}^{m,r}$, where $f \in \mathcal{S}(F)$, one computes easily

$$W\left(\begin{pmatrix} a & 0 \\ 0 & a^{-1} \end{pmatrix}(\lambda, 0, 0) \right) = \delta_m(a)|a|^{1/2} f\left(\frac{ra}{2m} + \lambda \right) \tag{5.11}$$

for $a \in F^*$, $\lambda \in F$. Thus we have the following description of the Kirillov space for representations of G^J.

5.7.3 Proposition. *Let π be a non-supercuspidal representation of G^J. Assuming that the $\psi^{n,r}$-Whittaker model with parameters $n, r \in F$ exists, the corresponding Kirillov space $\mathcal{K}(\pi)$ consists of all linear combinations of functions $h : F^* \times F \to \mathbb{C}$ of the following type:*

i) *If $\pi = \pi_{\chi,m}$ is a principal series representation with $\chi^2 \neq 1$, then*

$$h(a, \lambda) = |a|^{3/2} f(ra + 2m\lambda)(\chi(a)f_1(a) + \chi^{-1}(a)f_2(a))$$

$\left(f \in \mathcal{S}(F), \; f_1, f_2 \in \mathcal{S}(F)^+ \right).$

ii) *If $\pi = \pi_{\chi,m}$ is a principal series representation with $\chi^2 = 1$, then*

$$h(a, \lambda) = |a|^{3/2} f(ra + 2m\lambda)(\chi(a)f_1(a) + \chi^{-1}(a)v_F(a)f_2(a))$$

$\left(f \in \mathcal{S}(F), \; f_1, f_2 \in \mathcal{S}(F)^+ \right).$

iii) *If $\pi = \sigma_{\xi,m}$ is a special representation, then*

$$h(a, \lambda) = |a|^{3/2} f(ra + 2m\lambda)\chi(a)f_1(a)$$

$\left(f \in \mathcal{S}(F), \; f_1 \in \mathcal{S}(F)^+ \right).$

iv) *If $\pi = \sigma_{\xi,m}^+$ is a positive Weil representation, then*

$$h(a, \lambda) = |a|^{3/2} f(ra + 2m\lambda)\chi^{-1}(a)f_1(a)$$

$\left(f \in \mathcal{S}(F), \; f_1 \in \mathcal{S}(F)^+ \right).$

Caution: The function f appearing in these formulas is not quite the Schwartz function corresponding to W in the Whittaker model $\mathcal{W}_{SW}^{m,r}$. We have shifted the argument multiplicatively by $2m$ to avoid fractions, cf. (5.11).

5.8 Summary and Classification

We are almost ready to classify the irreducible, admissible representations of G^J. Only a few simple lemmas are still necessary.

5.8.1 Lemma. *Let $\varepsilon > 0$ and χ, χ' characters of F^* such that*

$$\chi(a) = \chi'(a) \qquad \text{for all } a \in F^* \text{ with } |a| < \varepsilon.$$

Then $\chi = \chi'$.

Proof: Let $a, a' \in F^*$ be arbitrary. The hypotheses implies the existence of some $N \in \mathbb{N}$ such that

$$\chi(\omega^n a) = \chi'(\omega^n a) \qquad \text{for all } n \geq N.$$

The number $(\chi(\omega)/\chi'(\omega))^n$ does not depend on n, hence $\chi(\omega) = \chi'(\omega)$. This in turn implies $\chi(a) = \chi'(a)$. $\qquad\square$

5.8.2 Lemma. *Let $c_1, c_2 \in \mathbb{C}$, $\varepsilon > 0$ and χ, χ' characters of F^*. If*

$$c_1 \chi(a) + c_2 \chi^{-1}(a) = \chi'(a) \qquad \text{for all } a \in F^* \text{ with } |a| < \varepsilon,$$

then $\chi' = \chi$ or $\chi' = \chi^{-1}$.

Proof: If

$$\chi(a) = \chi'(a) \qquad \text{for all } |a| < \varepsilon,$$

then Lemma 5.8.1 shows $\chi = \chi'$. Assume on the contrary that $\chi(a) \neq \chi'(a)$ for some $a \in F^*$ with $|a| < \varepsilon$. The hypotheses implies

$$c_1 \left(\frac{\chi(a)}{\chi'(a)} \right)^n + c_2 \left(\frac{\chi^{-1}(a)}{\chi'(a)} \right)^n = 1 \qquad \text{for all } n \geq 1.$$

It is a little exercise to deduce from this that $c_1 = 0$. Then it follows quickly that $\chi' = \chi^{-1}$. $\qquad\square$

5.8.3 Theorem. *Here is a complete list of the irreducible, admissible representations of the p-adic Jacobi group $G^J(F)$ with non-trivial central character ψ^m.*

   i) The principal series representations $\pi_{\chi,m}$.

   ii) The special representations $\sigma_{\xi,m}$.

   iii) The positive Weil representations $\sigma_{\xi,m}^{+}$.

   iv) The negative Weil representations $\sigma_{\xi,m}^{-}$.

   v) The supercuspidal representations not of the form $\sigma_{\xi,m}^{-}$ for some $\xi \in F^*$.

Between these representations there are exactly the following equivalences:

$$\pi_{\chi,m} \simeq \pi_{\chi^{-1},m}.$$

$$\sigma_{\xi,m} \simeq \sigma_{\xi',m} \qquad \Longleftrightarrow \qquad \xi F^{*2} = \xi' F^{*2}.$$

$$\sigma^{\pm}_{\xi,m} \simeq \sigma^{\pm}_{\xi',m} \qquad \Longleftrightarrow \qquad \xi F^{*2} = \xi' F^{*2}.$$

Proof: By Theorem 5.4.4 and the subrepresentation theorem 5.5.2 the given list is complete. The indicated equivalences hold by Corollary 5.6.6 resp. the trivial fact that the Hilbert symbol (a, ξ) depends only on ξ mod F^{*2}. We have to show that there are no other equivalences. So assume first that $\pi_{\chi,m} \simeq \pi_{\chi',m}$ for two principal series representations. We choose some Whittaker model for these and look at the corresponding Kirillov space. By the description in Proposition 5.7.3, there must exist $c, d \in \mathbb{C}$ such that

$$\chi'(a) = c\chi(a) + d\chi^{-1}(a)$$

for all $a \in F^*$ with small enough absolute value. Thus by the preceding lemma, we have indeed $\chi' = \chi$ or $\chi' = \chi^{-1}$. The same argument shows that a special or a Weil representation can not be equivalent to a principal series representation. It is even simpler to see that special and Weil representations are not equivalent. What remains to examine are the equivalences between two special resp. two Weil representations. So assume

$$\sigma_{\xi,m} \simeq \sigma_{\xi',m} \qquad \text{for some } \xi, \xi' \in F^*.$$

Again by looking at Kirillov models, we conclude that

$$c\chi(a) = \chi'(a)$$

for a constant $c \in \mathbb{C}$ and all small enough $a \in F^*$. Then c is easily seen to equal 1, hence $\chi = \chi'$ by Lemma 5.8.1. This means

$$(a, \xi) = (a, \xi') \qquad \text{for all } a \in F^*.$$

But the Hilbert symbol is non-degenerate in the sense that this equation implies $\xi F^{*2} = \xi' F^{*2}$. $\qquad\qquad\square$

Now we summarize the results so far obtained on the irreducible, admissible representations of G^J. From the Stone-von Neumann theorem it follows that each smooth representation π of G^J with central character ψ^m can be written as a tensor product

$$\pi = \tilde{\pi} \otimes \pi^m_{SW}$$

with a smooth representation $\tilde{\pi}$ of the metaplectic group Mp. Table 5.1 lists the relationships between π and $\tilde{\pi}$.

By the subrepresentation theorem 5.5.2 every non-supercuspidal irreducible admissible representation (with non-trivial central character) is induced. Table 5.2 summarizes the main properties of those induced representations which are not positive Weil representations.

π and $\tilde{\pi}$ are simultaneously	see
admissible	Proposition 5.1.2
irreducible	Lemma 5.4.3
supercuspidal	Proposition 5.5.1
induced	Theorem 5.4.2
pre-unitary	Proposition 5.9.1

Table 5.1: Relationship between π and $\tilde{\pi}$

name	principal series representation	special representation
symbol	$\pi_{\chi,m}$	$\sigma_{\xi,m}$
inducing character	$\chi^2 \neq |\,|, |\,|^{-1}$	$\chi = |\,|^{\pm 1/2}(\cdot,\xi),\ \xi \in F^*$
space	$\mathcal{B}^J_{\chi,m}$	subspace of $\mathcal{B}^J_{|\,|^{1/2}(\cdot,\xi),m}$
isomorphic to	$\pi_\chi \otimes \pi^m_{SW}$	$\sigma_\xi \otimes \pi^m_{SW}$
equivalences	$\pi_{\chi,m} \simeq \pi_{\chi^{-1},m}$	$\sigma_{\xi,m} \simeq \sigma_{\xi a^2,m}\ \forall a \in F^*$
$\mathcal{W}^{n,r}$ exists	if $N \neq 0$	if $-\xi N \in F^* \setminus F^{*2}$

Table 5.2: Non-archimedean principal series and special representations of G^J

name	positive and negative Weil representation
symbol	$\sigma^\pm_{\xi,m}$
isomorphic to	$\pi_w^{-m\xi\pm} \otimes \pi^m_{SW}$
equivalences	$\sigma^\pm_{\xi,m} \simeq \sigma^\pm_{\xi a^2,m}\ \forall a \in F^*$
$\mathcal{W}^{n,r}$ exists	if $-\xi N \in F^{*2}$

Table 5.3: Non-archimedean Weil representations of G^J

We have not classified the supercuspidal representations, but among them we found the negative Weil representations. Table 5.3 lists the basic properties of the positive and negative Weil representations.

5.9 Unitary representations

As usual, a smooth representation π of a $\mathfrak{p}$-adic group G on a complex vector space V is called **pre-unitary** if there exists a non-degenerate positive definite hermitian form $\langle\,,\,\rangle$ on V such that

$$\langle\pi(g)v, \pi(g)v'\rangle = \langle v, v'\rangle \qquad \text{for all } v, v' \in V, \, g \in G.$$

5.9.1 Proposition. Let $\tilde{\pi}$ be a smooth representation of Mp and $\pi = \tilde{\pi} \otimes \pi_{SW}^m$ the corresponding smooth representation of G^J. Then π is pre-unitary if and only if $\tilde{\pi}$ is pre-unitary.

Proof: Let V be the space of $\tilde{\pi}$ and $\mathcal{S}(F)$ the space of π_{SW}^m. We know that the ordinary L^2-scalar product

$$\langle f, f'\rangle_{\mathcal{S}(F)} = \int_F f(x)\overline{f'(x)}\, dx \tag{5.12}$$

on $\mathcal{S}(F)$ is invariant under π_{SW}^m. Consequently, if $\tilde{\pi}$ is pre-unitary with invariant positive definite form $\langle\,,\,\rangle_0$ on V, then one can define an invariant hermitian form $\langle\,,\,\rangle$ on $V \otimes \mathcal{S}(F)$ with the property

$$\langle\varphi \otimes f, \varphi' \otimes f'\rangle = \langle\varphi, \varphi'\rangle_0 \langle f, f'\rangle_{\mathcal{S}(F)}.$$

This form is indeed positive definite: Since $\mathcal{S}(F)$ is a separable pre-Hilbert space, it has a countable orthonormal basis. Write any given element $\Phi \neq 0$ of $V \otimes \mathcal{S}(F)$ in the form

$$\Phi = \sum_{\text{finite}} \varphi_i \otimes f_i$$

with f_i elements from such a basis and linear independent. Then obviously

$$\langle\Phi, \Phi\rangle = \sum\langle\varphi_i, \varphi_i\rangle_0 > 0.$$

Assume conversely that $\langle\,,\,\rangle$ is an invariant, positive definite hermitian form an $V \otimes \mathcal{S}(F)$. For any fixed $\varphi \in V$ the bilinear form

$$(f, f') \longmapsto \langle\varphi \otimes f, \varphi \otimes f'\rangle$$

is positive definite and H-invariant, and must consequently up to a constant coincide with the scalar product (5.12). Thus if we call this constant $c(\varphi)$, then

$$\langle\varphi \otimes f, \varphi \otimes f'\rangle = c(\varphi)\langle f, f'\rangle_{\mathcal{S}(F)}.$$

Now for $\varphi, \varphi' \in V$ define

$$\langle\varphi, \varphi'\rangle_0 := \frac{1}{4}\Big(c(\varphi + \varphi') - c(\varphi - \varphi') + ic(\varphi + i\varphi') - ic(\varphi - i\varphi')\Big).$$

A quick calculation yields

$$\langle \varphi, \varphi' \rangle_0 \langle f, f \rangle_{\mathcal{S}(F)} = \langle \varphi \otimes f, \varphi' \otimes f \rangle \qquad \text{for all } f \in \mathcal{S}(F).$$

So we have indeed constructed an invariant positive definite hermitian form on V. $\square$

Using this proposition together with well-known metaplectic results, we arrive at the following theorem.

5.9.2 Theorem. *Here is a complete list of all irreducible pre-unitary admissible representations of G^J with non-trivial central character ψ^m:*

i) *The principal series representations $\pi_{\chi,m}$ where χ is a unitary character of F^*.*

ii) *The principal series representations $\pi_{\chi,m}$ where χ is a real valued character of F^* such that*

$$|\chi| = |\,|^{\sigma} \qquad \text{with} \qquad -\frac{1}{2} < \sigma < \frac{1}{2}.$$

iii) *The special representations.*

iv) *The Weil representations.*

v) *The supercuspidal representations.*

Proof: We only sketch part of the proof, following the arguments in [Go2], p. 60. If an admissible representation is pre-unitary, then the complex conjugate representation is equivalent to the contragredient one. It is clear that the complex conjugate of $\mathcal{B}^J_{\chi,m}$ equals $\mathcal{B}^J_{\bar{\chi},-m}$ (consider the map $\Phi \mapsto \bar{\Phi}$), while the contragredient is given by Proposition 5.4.6. Therefore, if $\mathcal{B}^J_{\chi,m}$ is pre-unitary, then

$$\mathcal{B}^J_{\bar{\chi},-m} \simeq \mathcal{B}^J_{\chi^{-1},-m}.$$

Theorem 5.8.3 then yields the necessary condition

$$\bar{\chi} = \chi^{-1} \qquad \text{or} \qquad \bar{\chi} = \chi.$$

In the first case χ is a unitary character, and since we induce with the correct modular factor, these representations are indeed unitary. It remains to treat the second case, where χ is a real-valued character. This is a bit complicated, and we hope the reader will not worry when he is referred now to [Go2] for this. $\square$

5.9.3 Definition.

i) *The pre-unitary representations under point (ii) in the theorem above are called* **continuous series representations**.

ii) *The representations under point (iii) in the theorem are called* **complementary series representations**.

6

Spherical Representations

This section deals with a special class of local non-archimedean representations of G^J. The so-called spherical representations are characterized by the existence of a non-zero vector fixed by the compact-open subgroup $G^J(\mathcal{O})$, where $\mathcal{O}$ is the ring of integers in F. For global considerations it is always necessary to have sufficient information on spherical representations.

The main tool in studying them is the *spherical Hecke algebra*. Trying to reveal the structure of this important algebra one soon faces obstacles due to the fact that G^J is not reductive. Nevertheless, using again the trick of looking only at representations with fixed central character, one arrives for almost all places (which we call the "good" ones) at a Hecke algebra with a very nice and simple structure: It is a polynomial ring in one variable, just as for the group SL(2). These results were obtained earlier by Murase [Mu] (following unpublished ideas of Shintani) and Dulinski [Du].

Using this result, we will be able to decide in the good case which of the irreducible local representations classified in the last chapter are spherical (Theorem 6.3.10). The treatment of the good cases is enough for the global theory as it is developed later in this book, but it should be mentioned that some interesting and specific non-reductive phenomena occur at the bad places. These will be discussed in the forthcoming thesis by the second-named author.

Using the information on Whittaker models from the last chapter, we can compute the (up to scalars unique) invariant vector of a spherical representation in such a model. As a by-product we will obtain Hecke eigenvalues, Satake parameters, and local Euler factors. We remark that there exists related work by Sugano [Su]. Finally, we discuss the connection of our Hecke operators with the 'classic' ones introduced by Eichler and Zagier in [EZ].

6.1   The Hecke algebra of the Jacobi group

Throughout the chapter we choose an additive character ψ of the $\mathfrak{p}$-adic field F such that $\mathcal{O}$ is the biggest ideal of F on which ψ is trivial. This property is crucial for our treatment of the local Hecke algebra. In most cases, e.g. if F is some $\mathbb{Q}_p$, we can choose the additive standard character defined in Section 2.2.

In all that follows we fix

$$K^J := G^J(\mathcal{O}) = \mathrm{SL}(2, \mathcal{O}) \ltimes H(\mathcal{O})$$

as an open compact subgroup of G^J. But it should be noted that K^J is not maximal compact, and moreover, that G^J does not even contain any maximal compact subgroups. This is due to the fact that the center of G^J is isomorphic to F: Any compact subgroup of G^J can be made bigger by adding a large enough ideal in the center. However, it is certainly natural to use the above K^J, and our notion of spherical representation will always mean relative to K^J:

6.1.1 Definition. *A representation* $\pi : G^J \to \mathrm{GL}(V)$ *is called* **spherical** *if there exists a nonzero vector* $v \in V$ *which is fixed by* K^J:

$$\pi(k)v = v \qquad \text{for all } k \in K^J.$$

The most important tool when dealing with spherical representations is the **spherical Hecke algebra** $\mathcal{H}(G^J, K^J)$, which is by definition the vector space of locally constant K^J-biinvariant functions $\varphi : G^J \to \mathbb{C}$ with compact support, given the structure of an associative $\mathbb{C}$-algebra by the convolution product

$$(\varphi * \varphi')(x) = \int\limits_{G^J} \varphi(xy)\varphi'(y^{-1})\,dy \qquad (x \in F).$$

The connection with spherical representations is that if $\pi : G^J \to V$ is an admissible representation, and V^{K^J} denotes the (finite-dimensional) space of K^J-invariant vectors in V, then $\mathcal{H}(G^J, K^J)$ acts on V^{K^J} by the rule

$$\pi(\varphi)v = \int\limits_{G^J} \varphi(x)\pi(x)v\,dx.$$

It is well known that this establishes an injection

$$\begin{Bmatrix} \text{Irreducible, admissible, sphe-} \\ \text{rical representations of } G^J \end{Bmatrix} \hookrightarrow \begin{Bmatrix} \text{Irreducible, finite-dimensional,} \\ \text{smooth } \mathcal{H}(G^J, K^J)\text{-modules} \end{Bmatrix}.$$

The structure of $\mathcal{H}(G^J, K^J)$ is very complicated. For instance, in contrast to Hecke algebras for reductive groups, the Jacobi Hecke algebra is not commutative (see Section 5 in [Gr]), so that for an irreducible spherical representation of G^J one can in general not conclude for the one-dimensionality of the space of K^J-invariant vectors. However, the situation becomes considerably simpler if attention is drawn to representations with fixed non-trivial central character ψ^m. Such a representation being spherical certainly demands for

$$n := v(m) \geq 0,$$

because $\mathcal{O}$ is the biggest ideal of F on which our character ψ is trivial. We shall assume this in the sequel. After having fixed the central character of our representations, we can work with a simpler kind of Hecke algebra.

6.1.2 Definition. For $u \in F^*$ we define the **Hecke algebra with character**

$$\mathcal{H}(G^J, K^J, \psi^u)$$

to be the space of functions $\varphi : G^J \to \mathbb{C}$ which are compactly supported modulo Z, and which have the property

$$\varphi(kgk'z) = \psi^u(z)\varphi(g) \qquad \text{for all } g \in G^J, \ k, k' \in K^J, \ z \in Z.$$

$\mathcal{H}(G^J, K^J, \psi^u)$ is made into an associative algebra by the convolution product

$$(\varphi * \varphi')(x) = \int\limits_{G^J/Z} \varphi(xy)\varphi'(y^{-1}) \, dy \qquad (x \in F).$$

This sort of Hecke algebra is indeed simpler then $\mathcal{H}(G^J, K^J)$: There is a natural homomorphism of $\mathbb{C}$-algebras

$$\Xi_u : \mathcal{H}(G^J, K^J) \longrightarrow \mathcal{H}(G^J, K^J, \psi^u), \tag{6.1}$$

$$\varphi \longmapsto \left(x \mapsto \int\limits_Z \varphi(xz)\psi^{-u}(z) \, dz \right). \tag{6.2}$$

This map is not hard to recognize to be surjective, hence $\mathcal{H}(G^J, K^J, \psi^u)$ is a quotient of $\mathcal{H}(G^J, K^J)$. To determine the kernel of Ξ_u we follow [Du] and define for $g \in G^J$

$$\varphi_g = 1_{K^J g K^J}$$

and

$$Z(g) = \{z \in Z : K^J g K^J = K^J g z K^J\}.$$

6.1.3 Lemma. For every $g \in G$, the subgroup $Z(g)$ of Z depends only on the class of g in $K^J \backslash G^J / K^J Z$. We have

$$\Xi_u(\varphi_g) = 0 \quad \text{if and only if} \quad \psi^u\big|_{Z(g)} \neq 1.$$

Proof: This is a simple exercise. $\qquad\qquad\qquad\qquad\qquad\qquad\qquad\qquad\square$

Now, if π is a spherical representation of G^J on a vector space V with non-trivial central character ψ^m, then the operation of $\mathcal{H}(G^J, K^J)$ on V^{K^J} factors through $\mathcal{H}(G^J, K^J, \psi^{-m})$, meaning that $\mathcal{H}(G^J, K^J, \psi^{-m})$ operates on V^{K^J} by

$$\pi(\varphi)v = \int\limits_{G^J/Z} \varphi(x)\pi(x)v \, dx.$$

Just as before, this yields an injection

$$\left\{ \begin{array}{l} \text{Irreducible, admissible, sphe-} \\ \text{rical representations of } G^J \text{ with} \\ \text{central character } \psi^m \end{array} \right\} \hookrightarrow \left\{ \begin{array}{l} \text{Irreducible, finite-} \\ \text{dimensional, smooth} \\ \mathcal{H}(G^J, K^J, \psi^{-m})\text{-modules} \end{array} \right\}. \quad (6.3)$$

The above lemma suggests that as $n = v(m)$ gets bigger, the kernel of Ξ_{-m} gets smaller. This is not quite true, but nevertheless it turns out that the Hecke algebra $\mathcal{H}(G^J, K^J, \psi^{-m})$ gets more and more complicated with increasing n. We refer to

- $n = 0$ as the **good case**,
- $n = 1$ as the **almost good case**,
- $n \geq 2$ as the **bad cases**.

The reason for this is that for $n = 0$ the Hecke algebra $\mathcal{H}(G^J, K^J, \psi^{-m})$ is isomorphic to a polynomial ring, in particular commutative, for $n = 1$ it is not very far from that and still commutative, while in the other cases it fails to be commutative. We will prove at least the first statement in the sequel; for the second one cf. [Du].

6.2   Structure of the Hecke algebra in the good case

In this section we will assume

$$n = v(m) = 0,$$

i.e., we are in the good case, and under this assumption the structure of $\mathcal{H}(G^J, K^J, \psi^{-m})$ will be determined completely (ψ continues to be a character with conductor $\mathcal{O}$). For brevity Ξ is written instead of Ξ_{-m}. It will turn out that our Hecke algebra is isomorphic to the well known Hecke algebra for SL(2). The following lemma points in this direction.

6.2.1 Lemma. *Let*

$$g = \begin{pmatrix} \omega^{-\alpha} & 0 \\ 0 & \omega^{\alpha} \end{pmatrix} (\lambda\omega^{-\alpha}, \mu, 0) \in G^J$$

with $\alpha \in \mathbb{N}_0$ and $\lambda, \mu \in F$. If

$$\Xi(\varphi_g) \neq 0, \qquad then \qquad \lambda, \mu \in \mathcal{O}.$$

Proof: Assume $\beta := -v(\lambda) \geq 0$ and $\gamma := -v(\mu) \geq 0$. We have to show $\beta = \gamma = 0$. Write $\lambda = \lambda_0/\omega^\beta$, $\mu = \mu_0/\omega^\gamma$ with units $\lambda_0, \mu_0 \in \mathcal{O}^*$. Let $s \in \mathcal{O}$. The identity

$$\begin{pmatrix} 1 & s \\ 0 & 1 \end{pmatrix} g = g(0, \lambda\omega^\alpha s, \lambda^2 s) \begin{pmatrix} 1 & s\omega^{2\alpha} \\ 0 & 1 \end{pmatrix}$$

shows:

$$\lambda\omega^\alpha s \in \mathcal{O} \implies \lambda^2 s \in Z(g).$$

Hence, if $\Xi(\varphi_g) \neq 0$, then by Lemma 6.1.3 we have the implication

$$\lambda\omega^\alpha s \in \mathcal{O} \implies \psi^{-m}(\lambda^2 s) = 1. \tag{6.4}$$

<u>Case 1:</u> $\beta \geq \alpha$.

In this case we let $s = u\omega^{\beta-\alpha}$ with $u \in \mathcal{O}$ arbitrary. It follows that

$$\lambda\omega^\alpha s = u\omega^{\beta-\alpha}\omega^\alpha\lambda_0\omega^{-\beta} \in \mathcal{O},$$

thus in view of (6.4)

$$1 = \psi^{-m}(s\lambda^2) = \psi(u\omega^{\beta-\alpha}\lambda_0^2\omega^{-2\beta}) = \psi(u\omega^{-(\alpha+\beta)}\lambda_0^2).$$

This holds for each $u \in \mathcal{O}$, such that $\psi^{-m}\big|_{\omega^{-(\alpha+\beta)}\mathcal{O}} = 1$. By the hypothesis on $n = v(m)$, this implies $\alpha + \beta \leq 0$, hence $\alpha = \beta = 0$.

<u>Case 2:</u> $\beta < \alpha$.

Then we have $\lambda\omega^\alpha s = \lambda\omega^{\alpha-\beta}s \in \mathcal{O}$ for all $s \in \mathcal{O}$, such that

$$1 = \psi(s\lambda^2) = \psi(s\lambda_0^2\omega^{-2\beta}) \qquad \text{for all } s \in \mathcal{O}.$$

We conclude that $\beta \leq 0$.

The statement about μ is proved similarly, starting with the equation

$$g\begin{pmatrix} 1 & 0 \\ -s & 1 \end{pmatrix} = \begin{pmatrix} 1 & 0 \\ -s\omega^{2\alpha} & 1 \end{pmatrix}(-\mu\omega^\alpha s, 0, \mu^2 s). \qquad \square$$

6.2.2 Proposition. *The Hecke algebra $\mathcal{H}(G^J, K^J, \psi^{-m})$ in the good case has a vector space basis given by the elements*

$$T^J(\omega^\alpha) := \Xi\left(\text{char}\left(K^J \begin{pmatrix} \omega^\alpha & 0 \\ 0 & \omega^{-\alpha} \end{pmatrix} K^J\right)\right), \qquad \alpha \in \mathbb{N}_0.$$

Proof: The large Hecke algebra $\mathcal{H}(G^J, K^J)$ has a basis consisting of functions φ_g. In view of the Cartan decomposition for $\text{SL}(2, F)$

$$\text{SL}(2, F) = \coprod_{\alpha \geq 0} K \begin{pmatrix} \omega^\alpha & 0 \\ 0 & \omega^{-\alpha} \end{pmatrix} K = \coprod_{\alpha \geq 0} K \begin{pmatrix} \omega^{-\alpha} & 0 \\ 0 & \omega^\alpha \end{pmatrix} K, \tag{6.5}$$

every such g may be assumed to have the form

$$g = (\lambda, 0, 0)\begin{pmatrix} \omega^{-\alpha} & 0 \\ 0 & \omega^{\alpha} \end{pmatrix}(0, \mu, \kappa) = \begin{pmatrix} \omega^{-\alpha} & 0 \\ 0 & \omega^{\alpha} \end{pmatrix}(\lambda\omega^{-\alpha}, \mu, \kappa + \lambda\omega^{-\alpha}).$$

The preceding lemma then states that $\Xi(\varphi_g)$ vanishes unless $\lambda, \mu \in \mathcal{O}$. The central component κ is irrelevant, such that $\mathcal{H}(G^J, K^J, \psi^{-m})$ is indeed spanned by the $T^J(\omega^{\alpha})$. Considering the matrix part of the support of these elements, it is clear that they are linearly independent, because of (6.5). $\qquad\square$

We will now compute how the elements $T(\omega^{\alpha})$ multiply. This will be done by explicit coset decompositions. The following result for SL(2) is fundamental and can be checked by elementary calculations.

6.2.3 Lemma.

$$\mathrm{SL}(2, \mathcal{O})\begin{pmatrix} \omega^{\alpha} & 0 \\ 0 & \omega^{-\alpha} \end{pmatrix}\mathrm{SL}(2, \mathcal{O})$$

$$= \coprod_{f=0}^{2\alpha} \coprod_{\substack{u \in \mathcal{O}/\omega^f \\ (u, \omega^f, \omega^{2\alpha-f})=1}} \mathrm{SL}(2, \mathcal{O})\begin{pmatrix} \omega^{\alpha-f} & u\omega^{-\alpha} \\ 0 & \omega^{f-\alpha} \end{pmatrix}$$

$$= \mathrm{SL}(2, \mathcal{O})\begin{pmatrix} \omega^{\alpha} & 0 \\ 0 & \omega^{-\alpha} \end{pmatrix} \cup \coprod_{f=1}^{2\alpha-1} \coprod_{\substack{u \in \mathcal{O}/\omega^f \\ (u, \omega)=1}} \mathrm{SL}(2, \mathcal{O})\begin{pmatrix} \omega^{\alpha-f} & u\omega^{-\alpha} \\ 0 & \omega^{f-\alpha} \end{pmatrix}$$

$$\cup \coprod_{u \in \mathcal{O}/\omega^{2\alpha}} \mathrm{SL}(2, \mathcal{O})\begin{pmatrix} \omega^{-\alpha} & u\omega^{-\alpha} \\ 0 & \omega^{\alpha} \end{pmatrix}.$$

From this it is not very difficult, though a little bit lengthy, to prove the following decomposition. We leave the details to the reader (cf. also Lemma 2.8 in [Su])

6.2.4 Corollary.

$$K^J\begin{pmatrix} \omega^{\alpha} & 0 \\ 0 & \omega^{-\alpha} \end{pmatrix}K^J = \coprod_{\lambda \in \mathcal{O}/\omega^{\alpha}} K^J\begin{pmatrix} \omega^{\alpha} & 0 \\ 0 & \omega^{-\alpha} \end{pmatrix}(\lambda, 0, 0)$$

$$\cup \coprod_{f=1}^{2\alpha-1} \coprod_{\substack{u \in \mathcal{O}/\omega^{\alpha} \\ (u, \omega)=1}} \coprod_{\lambda \in \mathcal{O}/\omega^{\alpha}} K^J\begin{pmatrix} \omega^{\alpha-f} & u\omega^{-\alpha} \\ 0 & \omega^{f-\alpha} \end{pmatrix}(\lambda, 0, 0)$$

$$\cup \coprod_{u \in \mathcal{O}/\omega^{2\alpha}} \coprod_{\mu \in \mathcal{O}/\omega^{\alpha}} K^J\begin{pmatrix} \omega^{-\alpha} & u\omega^{-\alpha} \\ 0 & \omega^{\alpha} \end{pmatrix}(0, \mu, 0)$$

6.2.5 Proposition. In $\mathcal{H}(G^J, K^J, \psi^{-m})$, the elements $T^J(\omega^\alpha)$ multiply as follows (we are in the good case $n = v(m) = 0$):

$$T^J(\omega)^2 = T^J(\omega^2) + q^3 + q^2,$$

$$T^J(\omega)T^J(\omega^\alpha) = T^J(\omega^{\alpha+1}) + q^3 T^J(\omega^{\alpha-1}) \qquad \text{for } \alpha \geq 2.$$

Proof: We only prove the first statement and hope the reader thereby gets enough insight into the idea to treat the second one on his own. If φ_α denotes the characteristic function of

$$K^J \begin{pmatrix} \omega^\alpha & 0 \\ 0 & \omega^{-\alpha} \end{pmatrix} K^J,$$

then we have to prove that

$$\Xi(\varphi_1 * \varphi_1 - \varphi_2 - q^3 - q^2) = 0$$

(φ_0 is the identity in $\mathcal{H}(G^J, K^J)$). We will show that the support S of the function on the left-hand side is empty. This is equivalent to $p(S)$ being empty, where $p \colon G^J \to G$ denotes the natural projection. Now $p(S) \subset G$ is obviously left and right K-invariant; hence, in view of (6.5), it is enough to show that

$$\begin{pmatrix} \omega^\alpha & 0 \\ 0 & \omega^{-\alpha} \end{pmatrix} \notin p(S) \qquad \text{for all } \alpha \geq 0.$$

According to Corollary 6.2.4 we have

$$\varphi_1 = \sum_{\lambda \in \mathcal{O}/\omega} \operatorname{char}\left(K^J \begin{pmatrix} \omega & 0 \\ 0 & \omega^{-1} \end{pmatrix} (\lambda, 0, 0) \right)$$

$$+ \sum_{\substack{u \in \mathcal{O}/\omega \\ (u,\omega)=1}} \sum_{\lambda \in \mathcal{O}/\omega} \operatorname{char}\left(K^J \begin{pmatrix} 1 & u\omega^{-1} \\ 0 & 1 \end{pmatrix} (\lambda, 0, 0) \right)$$

$$+ \sum_{u \in \mathcal{O}/\omega^2} \sum_{\mu \in \mathcal{O}/\omega} \operatorname{char}\left(K^J \begin{pmatrix} \omega^{-1} & u\omega^{-1} \\ 0 & \omega \end{pmatrix} (0, \mu, 0) \right).$$

Convoluting this with itself, we see that in the projection of the support of $\varphi_1 * \varphi_1$ to G the only diagonal matrices that can appear are

$$\begin{pmatrix} 1 & 0 \\ 0 & 1 \end{pmatrix}, \quad \begin{pmatrix} \omega & 0 \\ 0 & \omega^{-1} \end{pmatrix}, \quad \text{and} \quad \begin{pmatrix} \omega^2 & 0 \\ 0 & \omega^{-2} \end{pmatrix}$$

(up to units). The projected support of φ_2 contains only $\begin{pmatrix} \omega^2 & 0 \\ 0 & \omega^{-2} \end{pmatrix}$, and the projected support of $q^3 - q^2$ contains only the identity matrix. So what we really have to prove is

$$\begin{pmatrix} \omega^\alpha & 0 \\ 0 & \omega^{-\alpha} \end{pmatrix} \notin p(S) \qquad \text{for } \alpha = 0, 1, 2.$$

$\boxed{1}$ We begin with $\alpha = 2$. Multiplying $\varphi_1 * \varphi_1$ out, only one of the three times three terms yields the matrix $\begin{pmatrix} \omega^2 & 0 \\ 0 & \omega^{-2} \end{pmatrix}$, namely

$$\left(\sum_{\lambda \in \mathcal{O}/\omega} \operatorname{char}\left(K^J \begin{pmatrix} \omega & 0 \\ 0 & \omega^{-1} \end{pmatrix} (\lambda, 0, 0) \right) \right)$$

$$* \left(\sum_{\lambda' \in \mathcal{O}/\omega} \operatorname{char}\left(K^J \begin{pmatrix} \omega & 0 \\ 0 & \omega^{-1} \end{pmatrix} (\lambda', 0, 0) \right) \right)$$

$$= \sum_{\lambda, \lambda' \in \mathcal{O}/\omega} \operatorname{char}\left(K^J \begin{pmatrix} \omega^2 & 0 \\ 0 & \omega^{-2} \end{pmatrix} (\lambda \omega + \lambda', 0, 0) \right)$$

$$= \sum_{\lambda \in \mathcal{O}/\omega^2} \operatorname{char}\left(K^J \begin{pmatrix} \omega^2 & 0 \\ 0 & \omega^{-2} \end{pmatrix} (\lambda, 0, 0) \right).$$

This is exactly the first term in the decomposition 6.2.4 for $\alpha = 2$. Hence the $\begin{pmatrix} \omega^2 & 0 \\ 0 & \omega^{-2} \end{pmatrix}$-parts of $\varphi_1 * \varphi_1$ and φ_2 cancel (yet in $\mathcal{H}(G^J, K^J)$, before projecting with Ξ).

$\boxed{2}$ Now we treat the case $\alpha = 1$. The matrix $\begin{pmatrix} \omega & 0 \\ 0 & \omega^{-1} \end{pmatrix}$ occurs neither in the projected support of φ_2, nor in the projected support of $q^3 - q^2$. So if our assertion is true, then the $\begin{pmatrix} \omega & 0 \\ 0 & \omega^{-1} \end{pmatrix}$-part of $\varphi_1 * \varphi_1$ should vanish under Ξ. Looking at the matrices we see quickly that there is exactly one of the nine terms that yields $\begin{pmatrix} \omega & 0 \\ 0 & \omega^{-1} \end{pmatrix}$, namely

$$\left(\sum_{\lambda \in \mathcal{O}/\omega} \operatorname{char}\left(K^J \begin{pmatrix} \omega & 0 \\ 0 & \omega^{-1} \end{pmatrix} (\lambda, 0, 0) \right) \right)$$

$$* \left(\sum_{\substack{u' \in \mathcal{O}/\omega \\ (u', \omega) = 1}} \sum_{\lambda' \in \mathcal{O}/\omega} \operatorname{char}\left(K^J \begin{pmatrix} 1 & u'\omega^{-1} \\ 0 & 1 \end{pmatrix} (\lambda', 0, 0) \right) \right)$$

$$= \sum_{\substack{u' \in \mathcal{O}/\omega \\ (u', \omega) = 1}} \sum_{\lambda, \lambda' \in \mathcal{O}/\omega} \operatorname{char}\left(K^J \begin{pmatrix} \omega & 0 \\ 0 & \omega^{-1} \end{pmatrix} (\lambda + \lambda', \lambda u \omega^{-1}, -\lambda \lambda' u \omega^{-1}) \right)$$

$$= \sum_{\substack{u' \in \mathcal{O}/\omega \\ (u', \omega) = 1}} \sum_{\lambda, \lambda' \in \mathcal{O}/\omega} \operatorname{char}\left(K^J \begin{pmatrix} \omega & 0 \\ 0 & \omega^{-1} \end{pmatrix} (\lambda', 0, \lambda^2 u \omega^{-1}) \right)$$

$$= \sum_{u \in \mathcal{O}/\omega} \sum_{\lambda, \lambda' \in \mathcal{O}/\omega} \mathrm{char}\left(K^J \begin{pmatrix} \omega & 0 \\ 0 & \omega^{-1} \end{pmatrix} (\lambda', 0, \lambda^2 u \omega^{-1}) \right)$$

$$- \sum_{\lambda, \lambda' \in \mathcal{O}/\omega} \mathrm{char}\left(K^J \begin{pmatrix} \omega & 0 \\ 0 & \omega^{-1} \end{pmatrix} (\lambda', 0, 0) \right). \tag{6.6}$$

The first summand goes under Ξ to

$$\sum_{u \in \mathcal{O}/\omega} \sum_{\lambda, \lambda' \in \mathcal{O}/\omega} \Xi\left(\mathrm{char}\left(K^J \begin{pmatrix} \omega & 0 \\ 0 & \omega^{-1} \end{pmatrix} (\lambda', 0, \lambda^2 u \omega^{-1}) \right) \right)$$

$$= \sum_{u \in \mathcal{O}/\omega} \sum_{\lambda, \lambda' \in \mathcal{O}/\omega} \psi^m(\lambda^2 u \omega^{-1}) \Xi\left(\mathrm{char}\left(K^J \begin{pmatrix} \omega & 0 \\ 0 & \omega^{-1} \end{pmatrix} (\lambda', 0, 0) \right) \right)$$

The sum

$$\sum_{u \in \mathcal{O}/\omega} \psi^m(\lambda^2 u \omega^{-1}) \qquad (\lambda \in \mathcal{O})$$

is non-zero if and only if λ is not a unit (here our assumption $v(m) = 0$ comes in), in which case it equals q. Hence the above expression gives

$$q \sum_{\lambda' \in \mathcal{O}/\omega} \Xi\left(\mathrm{char}\left(K^J \begin{pmatrix} \omega & 0 \\ 0 & \omega^{-1} \end{pmatrix} (\lambda', 0, 0) \right) \right)$$

This shows that (6.6) indeed projects to zero under Ξ.

$\boxed{3}$ Finally, we have to look at the cosets in $\varphi_1 * \varphi_1$ which contain in the SL(2)-part the identity matrix. These are exactly the following:

$$\left(\sum_{\lambda \in \mathcal{O}/\omega} \mathrm{char}\left(K^J \begin{pmatrix} \omega & 0 \\ 0 & \omega^{-1} \end{pmatrix} (\lambda, 0, 0) \right) \right) *$$

$$* \left(\sum_{u' \in \mathcal{O}/\omega^2} \sum_{\mu' \in \mathcal{O}/\omega} \mathrm{char}\left(K^J \begin{pmatrix} \omega^{-1} & u'\omega^{-1} \\ 0 & \omega \end{pmatrix} (0, \mu', 0) \right) \right)$$

$$+ \left(\sum_{\substack{u \in \mathcal{O}/\omega \\ (u, \omega) = 1}} \sum_{\lambda \in \mathcal{O}/\omega} \mathrm{char}\left(K^J \begin{pmatrix} 1 & u\omega^{-1} \\ 0 & 1 \end{pmatrix} (\lambda, 0, 0) \right) \right) *$$

$$* \left(\sum_{\substack{u' \in \mathcal{O}/\omega \\ u' = -u}} \sum_{\lambda' \in \mathcal{O}/\omega} \mathrm{char}\left(K^J \begin{pmatrix} 1 & u'\omega^{-1} \\ 0 & 1 \end{pmatrix} (\lambda', 0, 0) \right) \right)$$

$$+ \left(\sum_{\substack{u \in \mathcal{O}/\omega^2 \\ u=0}} \sum_{\mu \in \mathcal{O}/\omega} \mathrm{char}\left(K^J \begin{pmatrix} \omega^{-1} & u\omega^{-1} \\ 0 & \omega \end{pmatrix} (0, \mu, 0) \right) \right) *$$

$$* \left(\sum_{\lambda' \in \mathcal{O}/\omega} \mathrm{char}\left(K^J \begin{pmatrix} \omega & 0 \\ 0 & \omega^{-1} \end{pmatrix} (\lambda', 0, 0) \right) \right).$$

We leave it to the reader to prove by the methods already used in $\boxed{2}$ that the first summand projects under Ξ to q^3 (times the identity) and the other two summands add up to project to q^2. This finally completes the proof. $\square$

6.2.6 Corollary. *In the good case $n = 0$ the Hecke algebra $\mathcal{H}(G^J, K^J, \psi^{-m})$ is commutative and generated by $T^J(\omega)$.*

6.2.7 Corollary. *With X an indeterminate, the formal identity*

$$\sum_{\alpha=0}^{\infty} T^J(\omega^\alpha) X^\alpha = \frac{1 - q^2 X^2}{1 - T^J(\omega)X + q^3 X^2}$$

holds.

We go one step further and establish the **Satake isomorphism** for our Hecke algebra $\mathcal{H}(G^J, K^J, \psi^{-m})$. This could be done in a systematic way by computing integrals, as one does in the reductive case (cf., for instance, formula (19) in [Ca]). For such a treatment see Murase [Mu] or Dulinski [Du]. However, our situation is so simple that we can reach our aim faster by an ad hoc method.

Let W be the Weyl group of G^J, and let the non-trivial element of W operate on the polynomial ring $\mathbb{C}[X^{\pm 1}]$ by interchanging X and X^{-1}. The ring of invariants

$$\mathbb{C}[X^{\pm 1}]^W$$

is again a polynomial ring in the one variable $X + X^{-1}$.

6.2.8 Theorem. *There is an isomorphism of $\mathbb{C}$-algebras*

$$\mathcal{H}(G^J, K^J, \psi^{-m}) \xrightarrow{\sim} \mathbb{C}[X^{\pm 1}]^W$$

which for $\alpha > 0$ maps

$$T^J(\omega^\alpha) \longmapsto q^{\frac{3\alpha}{2}}(X^\alpha + X^{-\alpha}) + (1 - q^{-1})q^{\frac{3\alpha}{2}} \sum_{\substack{j=-\alpha+2 \\ j \equiv \alpha \bmod 2}}^{\alpha-2} X^j$$

Proof: It is clear that the indicated map is an isomorphism of vector spaces. The only thing to check is that the polynomials on the right fulfill the relations of the $T^J(\omega^\alpha)$ in Proposition 6.2.5. This is an elementary calculation. $\square$

The usual Satake isomorphism for SL(2) says

$$\mathcal{H}(G, K) \simeq \mathbb{C}[X^{\pm 1}]^W,$$

but the connection between $T^J(\omega^\alpha)$ and

$$T(\omega^\alpha) = \mathrm{char}\left(K \begin{pmatrix} \omega^\alpha & 0 \\ 0 & \omega^{-\alpha} \end{pmatrix} K\right) \in \mathcal{H}(G, K)$$

is not as simple as one might think. The latter element identifies with the polynomial

$$q^\alpha(X^\alpha + X^{-\alpha}) + q^\alpha(1 - q^{-1}) \sum_{j=-\alpha+1}^{\alpha-1} X^j,$$

and the multiplication law is

$$T(\omega)^2 = T(\omega^2) + (q - 1)T(\omega) + q^2 + q,$$

$$T(\omega)T(\omega^\alpha) = T(\omega^{\alpha+1}) + (q - 1)T(\omega^\alpha) + q^2 T(\omega^{\alpha-1}) \qquad (\alpha \geq 2).$$

If we identify Hecke operators with polynomials, then one can explicitly formulate a relation between Jacobi- and SL(2)-operators. But since this relation is not very enlightning and will be of no use to us, we only mention this here and leave it as an exercise to be done if necessary.
We also give the rationality theorem for SL(2),

$$\sum_{\alpha=0}^{\infty} T(\omega^\alpha) X^\alpha = \frac{1 + (q - 1)X - qX^2}{1 + (q - 1 - T(\omega))X + q^2 X^2},$$

which the reader might wish to compare with Corollary 6.2.7.

We can see from these formulas that the Hecke algebras of G^J and of SL(2), both being polynomial rings, are isomorphic, but not "canonical": The basic elements $T^J(\omega)$ and $T(\omega)$ correspond to different polynomials. This situation changes if we do not compare G^J with SL(2), but with PGL(2). Let G_1 denote the group $\mathrm{PGL}(2, F) \simeq \mathrm{SO}(3, F)$ and K_1 its maximal compact subgroup. The Satake isomorphism for the Hecke algebra of this pair is easily computed (see [Ca]) and yields

$$\mathcal{H}(G_1, K_1) \simeq \mathbb{C}[X^{\pm 1}]^W, \tag{6.7}$$

the same polynomial ring as before. But if now $T^{PGL}(\omega)$ denotes the basic Hecke operator, i.e., the characteristic function of the double coset

$$K_1 \left[\begin{pmatrix} \omega & 0 \\ 0 & 1 \end{pmatrix} \right] K_1,$$

then the isomorphism (6.7) sends $T^{PGL}(\omega)$ to the polynomial $q^{1/2}(X + X^{-1})$. Therefore, we have a "canonical" isomorphism

$$\mathcal{H}(G^J, K^J, \psi^{-m}) \xrightarrow{\sim} \mathcal{H}(G_1, K_1),$$
$$T^J(\omega) \longmapsto q T^{PGL}(\omega).$$

Via the well known connection between characters of the Hecke algebra and spherical representations (see Proposition 6.3.1 below) there should thus exist a "canonical" correspondence between spherical representations of $G^J(F)$ and $PGL(2, F)$. This is in fact true, and is a special case of a "lifting map" between automorphic representations of G^J and $PGL(2)$. More details on this subject can be found in [Sch2].

6.3 Spherical representations in the good case

In the previous section we have completely determined the structure of the local Hecke algebra $\mathcal{H}(G^J, K^J, \psi^{-m})$ in the good case. It turned out just to be a polynomial ring. In particular, the commutativity of this algebra allows us to conclude that all of its non-trivial finite dimensional irreducible representations are one dimensional, i.e., they are just characters (algebra homomorphisms $\mathcal{H}(G^J, K^J, \psi^{-m}) \to \mathbb{C}$). Furthermore, the commutativity allows for application of the business of spherical functions, just like in [Ca], 4.3, 4.4, and the conclusion is that the arrow in (6.3) is surjective. Summing up, we have the following result.

6.3.1 Proposition. *In the good case, there is a natural 1-1 correspondence*

$$\left\{ \begin{array}{l} \textit{Irreducible, admissible,} \\ \textit{spherical representa-} \\ \textit{tions of } G^J \textit{ with} \\ \textit{non-trivial central} \\ \textit{character } \psi^m \end{array} \right\} \longleftrightarrow \mathrm{Hom}_{\mathrm{Alg}}\Big(\mathcal{H}(G^J, K^J, \psi^{-m}), \mathbb{C} \Big). \tag{6.8}$$

Next we want to figure out which of the representations of G^J classified in Section 5.8 are spherical. This is easy for the metaplectic group, and we will use the canonical isomorphism from Theorem 5.4.2 to take the results over to G^J. To do that we need some information on whether the Schrödinger-Weil representation is spherical or not. Although we are mainly interested in the good case, we state Proposition 6.3.4 below in greater generality, because it will also be needed at another point. So for the moment we give up our assumption $n = v(m) = 0$, demand only $n \geq 0$, and define

$$
n_0 := \begin{cases} \dfrac{n}{2} & \text{if } v(m) \text{ even,} \\[2mm] \dfrac{n-1}{2} & \text{if } v(m) \text{ odd.} \end{cases}
$$

6.3.2 Lemma. *If $\mathcal{S}(F)$ denotes the standard space for the Schrödinger-Weil representation, then we have*

$$
\mathcal{S}(F)^{H(\mathcal{O})} = \left\{ f \in \mathcal{S}(F) : \operatorname{supp}(F) \subset \omega^{-v(2m)}\mathcal{O}, \ f \text{ is } \mathcal{O}\text{-invariant} \right\},
$$

$$
\mathcal{S}(F)^{N(\mathcal{O})} = \left\{ f \in \mathcal{S}(F) : \operatorname{supp}(F) \subset \omega^{-n_0}\mathcal{O} \right\}.
$$

Proof: We have

$$
\pi_s^m(\lambda, \mu, \kappa)f = f \qquad \text{for all } \lambda, \mu, \kappa \in \mathcal{O}
$$

if and only if

$$
\psi(2mx\mu)f(x + \lambda) = f(x) \qquad \text{for all } \lambda, \mu \in \mathcal{O}, \ x \in F.
$$

Hence $f \in \mathcal{S}(F)^{H(\mathcal{O})}$ must be $\mathcal{O}$-invariant. If $f(x) \neq 0$, then

$$
\psi(2mx\mu) = 1 \qquad \text{for all } \mu \in \mathcal{O}.
$$

From this it follows that $2mx \in \mathcal{O}$, i.e. $v(x) \geq -v(2m)$. This proves the first assertion. The second one is treated similarly. $\qquad\square$

In the sequel w denotes as usual the matrix $\begin{pmatrix} 0 & 1 \\ -1 & 0 \end{pmatrix}$.

6.3.3 Lemma. *For all $k \in \mathbb{Z}$ we have*

$$
\pi_w^m(w)\mathbf{1}_{\omega^k\mathcal{O}} = \gamma_m(1)q^{-k-v(2m)/2}\mathbf{1}_{\omega^{-k-v(2m)}\mathcal{O}}
$$

(this holds even for any $m \in F^*$).

Proof: If dy denotes the additive measure on F which gives $\mathcal{O}$ the volume 1, then Fourier transformation has to be normalized by

$$
\hat{f}(x) = q^{-v(2m)/2} \int_F \psi(2mxy)f(y) \, dy
$$

to make Fourier inversion hold. Hence

$$\left(\pi_W^m(w)\mathbf{1}_{\omega^k\mathcal{O}}\right)(x) = \gamma_m(1)q^{-v(2m)/2}\int_F \psi(2mxy)\mathbf{1}_{\omega^k\mathcal{O}}(y)\,dy$$

$$= \gamma_m(1)q^{-v(2m)/2}\int_{\omega^k\mathcal{O}}\psi(2mxy)\,dy.$$

The integral is non-zero exactly for $v(2mx) \geq -k$, in which case it equals q^{-k}.
$\square$

6.3.4 Proposition. *The Weil representation π_W^m contains a K-invariant vector if and only if F has odd residue characteristic and $v(2m)$ is even. If this is fulfilled, then the K-invariant vector is unique up to scalar multiples, and is even K^J-invariant. In the model $\mathcal{S}(F)$ it is given by*

$$\mathbf{1}_{\omega^{-v(m)/2}\mathcal{O}}.$$

Proof: If $n = v(m)$ is even and 2 is a unit in F, then $\mathbf{1}_{\omega^{-n/2}\mathcal{O}}$ is K^J-invariant by the preceding two lemmas (note that $\gamma_m(1) = 1$ by Lemma 5.3.2 v)). Assume conversely that $f \in \mathcal{S}(F)$ is K-invariant. By Lemma 6.3.2, for all $y, y' \in F$ with $v(y') \geq n_0 - n$ we have

$$f(y + y') = \left(\pi_W^m(w)f\right)(y + y')$$

$$= \gamma_{\psi^m}(1)\int_F \psi(2m(y + y')z)f(z)\,dz$$

$$= \gamma_{\psi^m}(1)\int_{\omega^{-n_0}\mathcal{O}}\psi(2myz)\psi(2my'z)f(z)\,dz$$

$$= \gamma_{\psi^m}(1)\int_{\omega^{-n_0}\mathcal{O}}\psi(2myz)f(z)\,dz$$

$$= \left(\pi_W^m(w)f\right)(y) = f(y),$$

i.e.

$$f \text{ is } \omega^{n_0-n}\mathcal{O} - \text{invariant.} \tag{6.9}$$

If n is odd, then from this condition and Lemma 6.3.2 it follows that $f = 0$. If n is even, then it follows from the same conditions that f is a multiple of $\mathbf{1}_{\omega^{-n/2}\mathcal{O}}$. Assuming it to be non-trivial, Lemma 6.3.3 for $k = n/2$ yields $v(2) = 0$, i.e., F is not an extension of $\mathbb{Q}_2$.
$\square$

Now we can make the connection between spherical representations of G^J and of the metaplectic group. Here we call a smooth representation $\tilde{\pi} : \mathrm{Mp} \to \mathrm{GL}(V)$ spherical if there exists a non-zero $v \in V$ with

$$\tilde{\pi}(k, 1)v = v \qquad \text{for all } k \in K.$$

6.3.5 Proposition. *Let $\tilde{\pi} : \text{Mp} \to \text{GL}(V)$ be an admissible representation of Mp and $\pi = \tilde{\pi} \otimes \pi_{SW}^m$ the corresponding admissible representation of G^J with central character ψ^m. Assume the good case $v(m) = 0$, and assume F has odd residue characteristic. Then π is spherical if and only if $\tilde{\pi}$ is spherical. In this case*

$$(V \otimes \mathcal{S}(F))^{K^J} = V^K \otimes \mathbb{C}1_{\mathcal{O}} \tag{6.10}$$

is one-dimensional.

Proof: Let

$$v = \sum \varphi_i \otimes f_i$$

a K^J-invariant vector in $V \otimes \mathcal{S}(F)$, where we may assume the φ_i to be linearly independent. The Heisenberg group acts only on the f_i, and from the linear independence of the φ_i we conclude that $f_i \in \mathcal{S}(F)^{H(\mathcal{O})}$ for all i. Lemma 6.3.2 then says that all f_i are multiples of $1_{\mathcal{O}}$. Thus v is a pure tensor

$$v = \varphi \otimes 1_{\mathcal{O}}.$$

Now $1_{\mathcal{O}}$ is K^J-invariant by Proposition 6.3.4. Hence letting K act on v we see that φ is K-invariant, i.e. $\tilde{\pi}$ is spherical. $\qquad\square$

Now it is our intention to figure out which of the representations of Mp are spherical. Remember the definition of the induced representations $\mathcal{B}_{\chi,-m}$ in Section 5.3.

6.3.6 Lemma. *The induced representation $\mathcal{B}_{\chi,-m}$ of Mp is spherical if and only if*

$$\chi(a) = \delta_{-m}(a) \qquad \text{for all } a \in \mathcal{O}^*. \tag{6.11}$$

If the residue characteristic of F is odd and $v(m)$ is even, then this is equivalent to χ being unramified. If $\mathcal{B}_{\chi,-m}$ is spherical, then the (up to scalars) unique K-invariant vector is given by

$$\varphi\left(\left(\begin{pmatrix} a & x \\ 0 & a^{-1} \end{pmatrix}, \varepsilon\right)(k,1)\right) = \varepsilon\delta_{-m}(a)|a|\chi(a) \tag{6.12}$$

for $a \in F^*$, $x \in F$, $\varepsilon \in \{\pm 1\}$, $k \in K$.

Proof: This is clear because of the Iwasawa decomposition

$$\text{Mp} = \tilde{B}K$$

(we have written K for the set of all $(k,1)$, $k \in K$): A K-invariant function $\varphi \in \mathcal{B}_{\chi,-m}$ can be well-defined by (6.12) if and only if the inducing character is trivial on $\tilde{B} \cap K$. This condition is expressed by (6.11). The second assertion follows by v) in Lemma 5.3.2. $\qquad\square$

This lemma already tells us exactly in which cases the principal series representations of Mp are spherical. We need two more lemmas to treat the special representation.

6.3.7 Lemma. *Let $G' \to \mathrm{GL}(V)$ be an admissible representation of a $\mathfrak{p}$-adic group G', and let $K' < G'$ be an open and compact subgroup. Then every K'-invariant subspace of V has a K'-invariant complement.*

Proof: Every K'-representation is completely reducible, as is well known. Hence our assertion would be clear if V were finite-dimensional. But we can reduce to this case by decomposing V into (finite-dimensional!) isotypic components.   □

6.3.8 Lemma. *If F has odd residue characteristic and*

$$(a, \xi) = 1 \qquad \text{for all } a \in \mathcal{O}^* \tag{6.13}$$

holds for some $\xi \in F^*$, then $v(\xi)$ is even.

Proof: It is well known that the Hilbert symbol is trivial on $\mathcal{O}^* \times \mathcal{O}^*$ if F has odd residue characteristic. Hence (6.13) is true if $v(\xi)$ is even. If it also holds for some ξ with $v(\xi)$ odd, then it would hold for all $\xi \in F$. This would imply

$$\mathcal{O}^* \subset F^{*2}.$$

But in odd residue characteristic it is also well known that $\mathcal{O}^{*2}$ has index 2 in $\mathcal{O}^*$.   □

6.3.9 Proposition. *Assume the residue characteristic of F is odd and that we are in the good case $v(m) = 0$. We use the notion of Definition 5.3.4.*

  i) *The principal series representation π_χ is spherical if and only if χ is unramified.*

  ii) *The special representation σ_ξ is never spherical.*

  iii) *The positive Weil representation $\pi_w^{-m\xi}$ is spherical if and only if $v(\xi)$ is even.*

Proof: i) is already contained in Lemma 6.3.6.   iii) follows from Proposition 6.3.4. For the special representation σ_ξ to be spherical, the character $\chi = |\,|^{1/2}(\cdot, \xi)$ must be unramified by Lemma 6.3.6, because σ_ξ is a subrepresentation of $\mathcal{B}_{\chi, -m}$. Thus

$$(a, \xi) = 1 \qquad \text{for all } a \in \mathcal{O}^*$$

must necessarily hold. By Lemma 6.3.8, the valuation $v(\xi)$ must then be even. But in this case the Weil representation $\pi_w^{-m\xi}$ is spherical. Lemma 6.3.7 then implies that σ_ξ is not spherical, since $\mathcal{B}_{\chi, -m}$ contains the trivial representation of K at most once.   □

Now we are ready to state the main result of this section.

6.3.10 Theorem. *Assume the good case* $v(m) = 0$, *and assume* F *has odd residue characteristic. Then the following is a complete list of the irreducible, admissible, spherical representations of* $G^J(F)$ *with central character* ψ^m.

> i) *The principal series representation* $\pi_{\chi,m}$ *with unramified* χ.
>
> ii) *The positive Weil representation* $\sigma_{\xi,m}^+$ *with* $v(\xi)$ *even.*

Proof: In view of Propositions 6.3.5 and 6.3.9 the only thing that remains to prove is that in the good case supercuspidal representations are not spherical. This will follow from the considerations in Section 6.5 (see Corollary 6.5.6). □

6.4 Spherical Whittaker functions

In Theorem 6.3.10 we have determined – under good conditions – which of the irreducible, admissible representations of G^J are spherical. Now we want to compute explicitly the (up to scalars unique) spherical vector in the Whittaker models for these representations. Before doing this we give the spherical vector in another model, namely in the induced model, where it is easy to see. Consider the space $\mathcal{B}_{\chi,m}^J$ of the induced representation. To determine the spherical vector (if it exists) in this model we take together the following four items:

- The isomorphism $\mathcal{B}_\chi \otimes \mathcal{S}(F) \simeq \mathcal{B}_{\chi,m}^J$ from Theorem 5.4.2.

- The spherical vector in $\mathcal{B}_\chi$ from Proposition 6.3.6.

- The spherical vector for π_{SW}^m in $\mathcal{S}(F)$ from Proposition 6.3.4

- The equation (6.10) in Proposition 6.3.5.

6.4.1 Proposition. *Assume the good case* $v(m) = 0$ *and that* F *is not an extension of* $\mathbb{Q}_2$. *Then the representation* $\mathcal{B}_{\chi,m}^J$ *is spherical if and only if* χ *is unramified, and in this case a spherical vector is given by*

$$\Phi\left(\begin{pmatrix} a & x \\ 0 & a^{-1} \end{pmatrix}(\lambda, \mu, \kappa)\right) = |a|^{3/2}\chi(a)\psi^m(\kappa + \lambda\mu)\mathbf{1}_{\mathcal{O}}(\lambda)$$

for all $a \in F^*$ *and* $x, \lambda, \mu, \kappa \in F$.

Proof: In view of the above remarks there remains only to compute the Whittaker function $W_{\mathbf{1}_{\mathcal{O}}}$ corresponding to the Schwartz function $\mathbf{1}_{\mathcal{O}}$ in the model $\mathcal{W}_{SW}^{m,0}$. By Theorem 5.2.4 we have $W_{\mathbf{1}_{\mathcal{O}}}(g) = (\pi_{SW}^m(g)\mathbf{1}_{\mathcal{O}})(0)$ for all $g \in G^J$. An easy computation gives

$$W_{\mathbf{1}_{\mathcal{O}}}\left(\begin{pmatrix} a & x \\ 0 & a^{-1} \end{pmatrix}(\lambda, \mu, \kappa)\right) = \delta_m(a)|a|^{1/2}\psi^m(\kappa + \lambda\mu)\mathbf{1}_{\mathcal{O}}(\lambda),$$

and the assertion follows.   □

We could try to compute the spherical vector in a $\psi^{n,r}$-Whittaker model by applying to this result the Whittaker functional given by (5.10). However, this is not so easily done. Instead we prefer another method which utilizes the Hecke algebra. Let π either be a spherical principal series representation or a spherical positive Weil representation, induced from the unramified character χ of F^* (Theorem 6.3.10). *In all that follows we restrict ourselves to the case*

$$v(2m) = 0,$$

i.e., F has odd residue characteristic, and we are in the good case. Assume we have a $\psi^{n,r}$-Whittaker model $\mathcal{W}$ of π, where

$$N := 4mn - r^2 \neq 0,$$

(Theorem 5.7.2) and denote by W the (up to scalars) unique non-trivial K^J-invariant Whittaker function. By the 'Iwasawa decomposition'

$$G = NAK$$

and the Whittaker transformation property, we know W completely if we know the values

$$W(\tilde{d}(a, \lambda)), \qquad a \in F^*, \ \lambda \in F,$$

where in this section we abbreviate

$$\tilde{d}(a, \lambda) := \begin{pmatrix} a & 0 \\ 0 & a^{-1} \end{pmatrix} (\lambda, 0, 0) \in G^J(F) \qquad\qquad \text{for } a \in F^*, \ \lambda \in F.$$

6.4.2 Lemma. *If $W(\tilde{d}(a, \lambda)) \neq 0$, then $Na^2 \in \mathcal{O}$ and $ra + 2m\lambda \in \mathcal{O}$.*

Proof: W is K^J-invariant, hence for all $x, \mu \in F$

$$
\begin{aligned}
W(\tilde{d}(a, \lambda)) &= W\left(\tilde{d}(a, \lambda) \begin{pmatrix} 1 & x \\ 0 & 1 \end{pmatrix} (0, \mu, 0)\right) \\
&= \psi\Big(x(na^2 + ra\lambda + m\lambda^2) + \mu(ra + 2m\lambda)\Big) W(\tilde{d}(a, \lambda))
\end{aligned}
$$

must hold. If $W(\tilde{d}(a, \lambda)) \neq 0$, then it follows that

$$\psi\Big(x(na^2 + ra\lambda + m\lambda^2) + \mu(ra + 2m\lambda)\Big) = 1 \qquad\qquad \text{for all } x, \mu \in \mathcal{O}.$$

But our choice of ψ then forces

$$na^2 + ra\lambda + m\lambda^2 \in \mathcal{O} \qquad \text{and} \qquad ra + 2m\lambda \in \mathcal{O}.$$

The assertion follows from the identity

$$(ra + 2m\lambda)^2 = 4m(na^2 + ra\lambda + m\lambda^2) - Na^2$$

and the hypothesis $v(2m) = 0$. $\qquad\qquad\qquad\qquad\qquad\qquad\qquad\qquad\qquad\qquad\square$

From Proposition 6.3.5 and the way Whittaker models for G^J-representations are built from Whittaker models for Mp-representations and for π_{sw}^m, we know that W must be of the form

$$W(\tilde{d}(a,\lambda)) = W_{sw}(\tilde{d}(a,\lambda))F(a),$$

with W_{sw} the spherical Whittaker function for π_{sw}^m (corresponding to the Schwartz function $\mathbf{1}_{\mathcal{O}} \in \mathcal{S}(F)$) and F a function in the Kirillov model of the Mp-representation $\tilde{\pi}$ corresponding to π. It is easy to compute W_{sw}: By Theorem 5.2.4 we have

$$
\begin{aligned}
W_{sw}(\tilde{d}(a,\lambda)) &= \left(\pi_{sw}^m(\tilde{d}(a,\lambda))\mathbf{1}_{\mathcal{O}}\right)\left(\frac{r}{2m}\right) \\
&= \left(\pi_w\begin{pmatrix} a & 0 \\ 0 & a^{-1} \end{pmatrix}\pi_s(\lambda,0,0)\mathbf{1}_{\mathcal{O}}\right)\left(\frac{r}{2m}\right) \\
&= \delta_m(a)|a|^{1/2}\left(\pi_s(\lambda,0,0)\mathbf{1}_{\mathcal{O}}\right)\left(\frac{ra}{2m}\right) \\
&= \delta_m(a)|a|^{1/2}\mathbf{1}_{\mathcal{O}}\left(\lambda+\frac{ra}{2m}\right) \\
&= \delta_m(a)|a|^{1/2}\mathbf{1}_{\mathcal{O}}(ra+2m\lambda).
\end{aligned}
$$

The last step is legitimized by the fact that $2m$ is a unit, and is carried out merely to avoid fractions. By our description of Kirillov models in Proposition 5.7.3 we arrive at the following cases.

- If $\pi = \pi_{\chi,m}$ is a spherical principal series representation with $\chi^2 \neq 1$, then there exist $f_1, f_2 \in \mathcal{S}(F)^+$ such that for all $a \in F^*$, $\lambda \in F$

$$W(\tilde{d}(a,\lambda)) = |a|^{3/2}\mathbf{1}_{\mathcal{O}}(ra+2m\lambda)\left(\chi(a)f_1(a) + \chi^{-1}(a)f_2(a)\right).$$

- If $\pi = \pi_{\chi,m}$ is a spherical principal series representation with $\chi^2 = 1$, then there exist $f_1, f_2 \in \mathcal{S}(F)^+$ such that for all $a \in F^*$, $\lambda \in F$

$$W(\tilde{d}(a,\lambda)) = |a|^{3/2}\mathbf{1}_{\mathcal{O}}(ra+2m\lambda)\left(\chi(a)f_1(a) + \chi^{-1}(a)v(a)f_2(a)\right).$$

- If $\pi = \sigma_{\xi,m}^+$ is a spherical positive Weil representation, then there exists $f \in \mathcal{S}(F)^+$ such that for all $a \in F^*$, $\lambda \in F$

$$W(\tilde{d}(a,\lambda)) = |a|^{3/2}\mathbf{1}_{\mathcal{O}}(ra+2m\lambda)\chi^{-1}(a)f(a).$$

Notice that there are exactly two spherical principal series representations $\pi_{\chi,m}$ with $\chi^2 = 1$, according to the two unramified characters of F^* characterized by $\chi(\omega) = 1$ resp. $\chi(\omega) = -1$. By Theorem 6.3.10 there are also exactly two spherical Weil representations, because $\mathcal{O}^{*2}$ has index 2 in $\mathcal{O}^*$. They come from the unramified characters χ given by $\chi(\omega) = q^{-1/2}$ resp. $\chi(\omega) = -q^{-1/2}$.

6.4.3 Lemma. *The functions $f_1, f_2, f \in \mathcal{S}(F)^+$ in the above description of W may be assumed to be multiplicatively $\mathcal{O}^*$-invariant, i.e.*

$$f(aa') = f(a) \qquad \text{for all } a \in F^*, \, a' \in \mathcal{O}^*,$$

and similarly for f_1, f_2.

Proof: We prove this for the principal series representations with $\chi^2 \neq 1$. For any $a \in F^*$, $\lambda \in F$ and $a' \in \mathcal{O}^*$

$$W(\tilde{d}(a, \lambda)\tilde{d}(a', 0)) = W(\tilde{d}(a, \lambda)),$$

holds, which leads to

$$\chi(a)f_1(aa') + \chi^{-1}(a)f_2(aa') = \chi(a)f_1(a) + \chi^{-1}(a)f_2(a) \qquad (6.14)$$

for $a' \in \mathcal{O}^*$. Define for $i = 1, 2$

$$\tilde{f}_i(a) := \int_{\mathcal{O}^*} f_i(aa') \, d^*a', \qquad a \in F,$$

where d^*a' denotes multiplicative Haar measure. Then $\tilde{f}_i$ is a (multiplicatively) $\mathcal{O}^*$-invariant function in $\mathcal{S}(F)^+$. Because of

$$\chi(a)\tilde{f}_1(a) + \chi^{-1}(a)\tilde{f}_2(a) = \int_{\mathcal{O}^*} \left(\chi(a)f_1(aa') + \chi^{-1}(a)f_2(aa') \right) d^*a'$$

$$\overset{(6.14)}{=} \int_{\mathcal{O}^*} \left(\chi(a)f_1(a) + \chi^{-1}(a)f_2(a) \right) d^*a'$$

$$= \quad \chi(a)f_1(a) + \chi^{-1}(a)f_2(a) = W(\tilde{d}(a, \lambda)),$$

f_1 and f_2 may be replaced by $\tilde{f}_1$ and $\tilde{f}_2$. $\square$

In the following we treat only the case of principal series representations with $\chi^2 \neq 1$, and remark that the other cases can be treated analogously. According to the preceding lemma f_1 and f_2 can be written in the following form:

$$f_1 = \sum_{i \in \mathbb{Z}} b_i \, \mathbf{1}_{\omega^i \mathcal{O}^*}, \qquad b_i \in \mathbb{C},$$

$$f_2 = \sum_{i \in \mathbb{Z}} c_i \, \mathbf{1}_{\omega^i \mathcal{O}^*}, \qquad c_i \in \mathbb{C}.$$

The spherical function then reads

$$W(\tilde{d}(a, \lambda)) = q^{-3i/2} \mathbf{1}_{\mathcal{O}}(ra + 2m\lambda)\left(\chi(\omega)^i b_i + \chi(\omega)^{-i} c_i \right), \qquad i = v(a),$$

and it remains to determine the numbers b_i and c_i. Now introduce the integer

$$
l := \begin{cases} -\frac{1}{2}v(N) & \text{if } v(N) \text{ even,} \\ -\frac{1}{2}(v(N) - 1) & \text{if } v(N) \text{ odd.} \end{cases} \tag{6.15}
$$

We know from Lemma 6.4.2 that $W(\tilde{d}(a, \lambda))$ vanishes if $v(Na^2) < 0$, i.e. if $v(a) < l$. So it may be assumed that

$$
b_i = c_i = 0 \qquad \text{for } i < l.
$$

Furthermore, as f_1 and f_2 are constant on a neighbourhood of 0, the b_i and c_i become constant for large i:

$$
b_i = b \in \mathbb{C} \quad \forall\, i \gg 0, \qquad\qquad c_i = c \in \mathbb{C} \quad \forall\, i \gg 0. \tag{6.16}
$$

Finally, we can change finitely many of the c_i arbitrarily when the corresponding b_i are adjusted correctly: Just replace

$$
c_i \mapsto c_i + x, \qquad b_i \mapsto b_i - \chi(\omega)^{-2i} i^\varepsilon x, \qquad\qquad x \in F \text{ arbitrary.}
$$

Taking these facts together, the c_i may be assumed to look like this:

$$
c_i = \begin{cases} 0 & \text{for } i < l, \\ c & \text{for } i \geq l. \end{cases}
$$

The numbers

$$
W_i := W\!\left(\tilde{d}\!\left(\omega^i, -\frac{r}{2m}\omega^i\right)\right) = q^{-3i/2}\left(\chi(\omega)^i b_i + \chi(\omega)^{-i} c\right), \qquad i \geq l,
$$

will play the key role in determining c, the b_i, and the eigenvalue α of W under the Hecke operator $T^J(\omega) \in \mathcal{H}(G^J, K^J, \psi^{-m})$. This is because of the following decisive lemma.

6.4.4 Lemma. *The W_i fulfill the recursion formula*

$$
\left(\alpha - \left(\frac{-N\omega^{2i}}{\omega}\right)q\right) W_i = W_{i-1} + q^3 W_{i+1}, \qquad\qquad \text{for all } i \geq l,
$$

where $\alpha \in \mathbb{C}$ is the Hecke eigenvalue of W, defined by

$$
T^J(\omega)W = \alpha W,
$$

and where W_{l-1} is meant to be zero.

The symbol $\left(\frac{\cdot}{\omega}\right)$ used here is defined by

$$
\left(\frac{x}{\omega}\right) = \begin{cases} 0 & \text{if } x \in \omega\mathcal{O}, \\ 1 & \text{if } x \in \mathcal{O}^*,\ \bar{x} \in (\mathcal{O}/\omega)^2, \\ -1 & \text{if } x \in \mathcal{O}^*,\ \bar{x} \notin (\mathcal{O}/\omega)^2 \end{cases} \tag{6.17}
$$

for $x \in \mathcal{O}$. The proof of this lemma will be postponed to the end of the section. In the meantime we use it to compute α.

6.4.5 Lemma. *The Hecke eigenvalue α of the spherical Whittaker function $W \in \mathcal{W}^{n,r}$ is in all of the above three cases given by*

$$\alpha = q^{3/2}\Big(\chi(\omega) + \chi(\omega)^{-1}\Big).$$

Proof: We only treat the case of the principal series representation $\pi_{\chi,m}$ with $\chi^2 \neq 1$, the other ones being less complicated.

Case 1: $c = 0$.
Then from the recursion formula of the previous lemma it follows that

$$\alpha\chi(\omega)^i b_i = q^{3/2}\Big(\chi(\omega)^{i-1}b_{i-1} + \chi(\omega)^{i+1}b_{i+1}\Big) \qquad \text{for all } i \geq l+1. \tag{6.18}$$

If b (defined in (6.16)) were 0, one could conclude from this that all the b_i and thus also W would vanish. So we have $b \neq 0$, and for large enough i one reads off from (6.18) that α has the desired value.

Case 2: $c \neq 0$.
The recursion formula for large enough i reads

$$\alpha\Big(\chi(\omega)^i b + i^\varepsilon \chi(\omega)^{-i}c\Big) = q^{3/2}\Big(\chi(\omega)^{i-1}b +$$
$$+ (i-1)^\varepsilon \chi(\omega)^{-(i-1)}c + \chi(\omega)^{i+1}b + (i+1)^\varepsilon \chi(\omega)^{-(i+1)}c\Big),$$

which can be rewritten as

$$\chi(\omega)^{2i}b\Big(\alpha - q^{3/2}(\chi(\omega) + \chi(\omega)^{-1})\Big) = c\Big(q^{3/2}(\chi(\omega) + \chi(\omega)^{-1}) - \alpha\Big).$$

The right side does not depend on i, thus for the left one to do the same, it is necessary that α has the desired value. $\qquad\square$

Now the final result can be stated.

6.4.6 Theorem. *Assume $v(2m) = 0$ and let $\mathcal{W}^{n,r}$ the Whittaker model with parameters $n, r \in F$ for the irreducible, spherical, representation π of G^J. Let $W \in \mathcal{W}^{n,r}$ be the (up to scalars) unique K^J-invariant function.*

i) *If $\pi = \pi_{\chi,m}$ is a principal series representation with $\chi^2 \neq |\ |, 1$, then*

$$W(\tilde{d}(a, \lambda)) = |a|^{3/2}\mathrm{char}_{\mathcal{O}}(ra + 2m\lambda)\Big(\chi(a)b + \chi^{-1}(a)c\Big)\mathbf{1}_{\mathcal{O}}(Na^2)$$

for all $a \in F^*$, $\lambda \in F$, where

$$b = q^{3l/2}\frac{\chi(\omega)^{1-l} - \beta}{\chi(\omega) - \chi(\omega)^{-1}}, \qquad c = q^{3l/2}\frac{\beta - \chi(\omega)^{l-1}}{\chi(\omega) - \chi(\omega)^{-1}}.$$

ii) If $\pi = \pi_{\chi,m}$ is a principal series representation with $\chi^2 = 1$, then

$$W(\tilde{d}(a, \lambda)) = |a|^{3/2}\mathrm{char}_\mathcal{O}(ra + 2m\lambda)\Big(\chi(a)b + v(a)\chi^{-1}(a)c\Big)\mathbf{1}_\mathcal{O}(Na^2)$$

for all $a \in F^*$, $\lambda \in F$, where

$$b = q^{3l/2}\chi(\omega)^{l+1}\left(1 - l\chi(\omega) - l\beta\right), \qquad c = q^{3l/2}\chi(\omega)^{l+1}\left(\chi(\omega) - \beta\right)$$

iii) If $\pi = \sigma^+_{m,\xi}$ is a positive Weil representation, then

$$W(\tilde{d}(a, \lambda)) = |a|^{3/2}\mathrm{char}_\mathcal{O}(ra + 2m\lambda)\chi^{-1}(a)\mathrm{char}_\mathcal{O}(Na^2)$$

for all $a \in F^*$, $\lambda \in F$.

Here we have $\beta = \left(\frac{-N\omega^{2l}}{\omega}\right)q^{-1/2}$, $N = 4mn - r^2$, l is defined in (6.15), and the symbol $\left(\frac{\cdot}{\omega}\right)$ in (6.17). The Hecke eigenvalue of W is in all cases

$$\alpha = q^{3/2}\Big(\chi(\omega) + \chi(\omega)^{-1}\Big).$$

Proof: Again only the more difficult case of a principal series representation is treated. Assume $\chi^2 \neq 1$. Under our previous assumption on the c_i, the recursion formula for the W_i from Lemma 6.4.4 with the value of α from Lemma 6.4.5 takes the form

$$\Big(\chi(\omega) + \chi(\omega)^{-1}\Big)b_i = \chi(\omega)^{-1}b_{i-1} + \chi(\omega)b_{i+1} \qquad \text{for all } i \geq l+1.$$

The b_i becoming constant, equal to b, for large i, one deduces from this that all the b_i for $i \geq l$ take on the same value b. So we have shown that W is indeed of the form described in the theorem, and it remains to determine the constants b and c. For this purpose we normalize W as follows:

$$1 \overset{!}{=} W_l = q^{-3l/2}\Big(\chi(\omega)^l b + \chi(\omega)^{-l}c\Big).$$

As a second equation we use the recursion formula for $i = l$:

$$\left(\alpha - \left(\frac{-N\omega^{2l}}{\omega}\right)q\right) = q^3 W_{l+1} = q^3 q^{-3(l+1)/2}\Big(\chi(\omega)^{l+1}b + \chi(\omega)^{-(l+1)}c\Big).$$

Substituting the known value for α from Lemma 6.4.5, these two equations lead to the values for b and c given in the theorem. □

There still remains to prove Lemma 6.4.4. We show the following more general statement.

6.4.7 Proposition. *The action of $T^J(\omega)$ on a spherical function $W \in \mathcal{W}^{n,r}$ is given by*

$$
(T^J(\omega)W)(\tilde{d}(a,\lambda)) \;=\; W\left(\tilde{d}\left(a\omega^{-1}, -\frac{r}{2m}a\omega^{-1}\right)\right)
$$
$$
+ \left(\frac{-Na^2}{\omega}\right) q\, W(\tilde{d}(a,\lambda)) + q^3\, W(\tilde{d}(a\omega, \lambda\omega)),
$$

if $ra + 2m\lambda \in \mathcal{O}$ and $Na^2 \in \mathcal{O}$, and

$$
(T^J(\omega)W)(\tilde{d}(a,\lambda)) = 0
$$

otherwise.

Proof: Let $ra + 2m\lambda \in \mathcal{O}$ and $Na^2 \in \mathcal{O}$; the other case is clear by Lemma 6.4.2. The operator $T^J(\omega)$ acts by convolution with the characteristic function of the double coset $K^J \begin{pmatrix} \omega & 0 \\ 0 & \omega^{-1} \end{pmatrix} K^J$. From Lemma 6.2.4 it is easy to deduce

$$
K^J \begin{pmatrix} \omega^{-1} & 0 \\ 0 & \omega \end{pmatrix} K^J = \coprod_{\lambda \in \mathcal{O}/\omega} \left(\begin{pmatrix} \omega^{-1} & 0 \\ 0 & \omega \end{pmatrix}, \lambda\omega^{-1}, 0, 0 \right) K^J
$$
$$
\cup \coprod_{\substack{u \in \mathcal{O}/\omega \\ (u,\omega)=1}} \coprod_{\mu \in \mathcal{O}/\omega} \left(\begin{pmatrix} 1 & u\omega^{-1} \\ 0 & 1 \end{pmatrix}, 0, u\mu\omega^{-1}, u\mu^2\omega^{-1} \right) K^J
$$
$$
\cup \coprod_{u \in \mathcal{O}/\omega^2} \coprod_{\mu \in \mathcal{O}/\omega} \left(\begin{pmatrix} \omega & u\omega^{-1} \\ 0 & \omega^{-1} \end{pmatrix}, 0, \mu\omega^{-1}, 0 \right) K^J.
$$

Hence

$$
(T^J(\omega)W)(\tilde{d}(a,\lambda)) = \sum_{\lambda' \in \mathcal{O}/\omega} W\left(\tilde{d}(a,\lambda) \left(\begin{pmatrix} \omega^{-1} & 0 \\ 0 & \omega \end{pmatrix}, \lambda'\omega^{-1}, 0, 0 \right) \right)
$$
$$
+ \sum_{\substack{u \in \mathcal{O}/\omega \\ (u,\omega)=1}} \sum_{\mu \in \mathcal{O}/\omega} W\left(\tilde{d}(a,\lambda) \left(\begin{pmatrix} 1 & u\omega^{-1} \\ 0 & 1 \end{pmatrix}, 0, u\mu\omega^{-1}, u\mu^2\omega^{-1} \right) \right)
$$
$$
+ \sum_{u \in \mathcal{O}/\omega^2} \sum_{\mu \in \mathcal{O}/\omega} W\left(\tilde{d}(a,\lambda) \left(\begin{pmatrix} \omega & u\omega^{-1} \\ 0 & \omega^{-1} \end{pmatrix}, 0, \mu\omega^{-1}, 0 \right) \right)
$$

$$
= \sum_{\lambda' \in \mathcal{O}/\omega} W\left(\tilde{d}(a\omega^{-1}, \lambda\omega^{-1} + \lambda'\omega^{-1})) \right) \tag{6.19}
$$

$$+ \sum_{\substack{u \in \mathcal{O}/\omega \\ (u,\omega)=1}} \sum_{\mu \in \mathcal{O}/\omega} W\Big(n(u\omega^{-1}a^2, u\omega^{-1}(\mu a + \lambda a))$$

$$\tilde{d}(a,\lambda)(u\omega^{-1}(\mu^2 + 2\lambda\mu + \lambda^2))\Big) \quad (6.20)$$

$$+ \sum_{u \in \mathcal{O}/\omega^2} \sum_{\mu \in \mathcal{O}/\omega} W\Big(n(a^2 u, a(\lambda u + \mu))\tilde{d}(a\omega, \lambda\omega)(\lambda^2 u + 2\lambda\mu)\Big), \quad (6.21)$$

where we have used the abbreviation

$$n(x,\mu) = \begin{pmatrix} 1 & x \\ 0 & 1 \end{pmatrix} (0,\mu,0) \qquad \text{for } x, \mu \in F.$$

The expressions (6.19), (6.20), (6.21) will be shown to equal the terms given in the proposition. First we have

$$v(ra\omega^{-1} + 2m(\lambda\omega^{-1} + \lambda'\omega^{-1})) = v(\omega^{-1}(ra + 2m\lambda + 2m\lambda')) \geq 0$$
$$\Leftrightarrow \quad ra + 2m\lambda + 2m\lambda' \in \omega\mathcal{O}$$
$$\Leftrightarrow \quad \lambda' \equiv -\Big(\lambda + \frac{ra}{2m}\Big) \bmod \omega,$$

so that by Lemma 6.4.2 the sum (6.19) reduces to

$$\sum_{\lambda' \in \mathcal{O}/\omega} W\Big(\tilde{d}(a\omega^{-1}, \lambda\omega^{-1} + \lambda'\omega^{-1})\Big)$$

$$= W\Big(\tilde{d}\Big(a\omega^{-1}, \lambda\omega^{-1} - \Big(\lambda + \frac{ra}{2m}\Big)\omega^{-1}\Big)\Big)$$

$$= W\Big(\tilde{d}\Big(a\omega^{-1}, -\frac{r}{2m}a\omega^{-1}\Big)\Big). \quad (6.22)$$

For (6.20) we have by the Whittaker transformation property

$$(6.20) = b(N,a)W(\tilde{d}(a,\lambda))$$

with

$$b(N,a) := \sum_{\substack{u \in \mathcal{O}/\omega \\ (u,\omega)=1}} \sum_{\mu \in \mathcal{O}/\omega}$$

$$\psi\Big(u\omega^{-1}(na^2 + ra\lambda + m\lambda^2 + \mu(ra + 2m\lambda) + m\mu^2)\Big). \quad (6.23)$$

These numbers are computed in the next lemma and give the desired values (at the moment they should perhaps be called $b(n,r,m,a,\lambda)$, because it is not at all clear that they depend only on N and a). Finally we have by the Whittaker transformation property

$$(6.21) = \sum_{u \in \mathcal{O}/\omega^2} \sum_{\mu \in \mathcal{O}/\omega} \psi\Big(u(na^2 + ra\lambda + m\lambda^2) + \mu(ra + 2m\lambda)\Big)$$

$$W(\tilde{d}(a\omega, \lambda\omega))$$

$$= q^3 W(\tilde{d}(a\omega, \lambda\omega)). \qquad \square$$

It remains to prove the following lemma, which is also valid without our general assumption that F has odd residue characteristic.

6.4.8 Lemma. For $a \in F^*$ and $\lambda \in F$ with

$$ra + 2m\lambda \in \mathcal{O} \quad \text{and} \quad na^2 + ra\lambda + m\lambda^2 \in \mathcal{O}$$

one has

$$b(N, a) = \begin{cases} \left(\dfrac{-Na^2}{\omega}\right) q & \text{if } q \text{ odd,} \\ 0 & \text{if } q \text{ even,} \end{cases}$$

where $N = 4mn - r^2$.

Proof: A simple calculation shows

$$b(N, a) = \sum_{\substack{u \in \mathcal{O}/\omega \\ (u,\omega)=1}} \sum_{\mu \in \mathcal{O}/\omega} \psi\left(\frac{1}{4m}u\omega^{-1}\left(Na^2 + (ra + 2m\lambda + \mu)^2\right)\right).$$

Because of $ra + 2m\lambda \in \mathcal{O}$ this simplifies to

$$b(N, a) = \sum_{\substack{u \in \mathcal{O}/\omega \\ (u,\omega)=1}} \sum_{\mu \in \mathcal{O}/\omega} \psi\left(\frac{1}{4m}u\omega^{-1}\left(Na^2 + \mu^2\right)\right).$$

First let q be odd.

Case 1: $Na^2 \in \omega\mathcal{O}$

Then one has

$$\begin{aligned}
b(N, a) &= \sum_{\substack{u \in \mathcal{O}/\omega \\ (u,\omega)=1}} \sum_{\mu \in \mathcal{O}/\omega} \psi\left(\frac{1}{4m}u\omega^{-1}\mu^2\right) \\
&= q - 1 + \sum_{\substack{u \in \mathcal{O}/\omega \\ (u,\omega)=1}} \sum_{\substack{\mu \in \mathcal{O}/\omega \\ (u,\omega)=1}} \psi\left(\frac{1}{4m}u\omega^{-1}\right) \\
&= q - 1 + \sum_{\substack{\mu \in \mathcal{O}/\omega \\ (u,\omega)=1}} (-1) = 0.
\end{aligned}$$

Case 2: $Na^2 \in \mathcal{O}^*$, $\overline{-Na^2} \notin (\mathcal{O}/\omega)^2$. (The bar denotes the residue class in $\mathcal{O}/\omega\mathcal{O}$.)

Then for all $\mu \in \mathcal{O}$ we have $\overline{-Na^2} \neq \bar{\mu}^2$, hence $\mu^2 + Na^2 \in \mathcal{O}^*$, and this implies

$$b(N, a) = \sum_{\mu \in \mathcal{O}/\omega} (-1) = -q.$$

Case 3: $Na^2 \in \mathcal{O}^*$, $\overline{-Na^2} \in (\mathcal{O}/\omega)^2$.

When μ runs through the cyclic group $(\mathcal{O}/\omega)^*$, then μ^2 runs exactly twice through the group $(\mathcal{O}/\omega)^{*2}$. So we have

$$b(N, a) = \sum_{\substack{u \in \mathcal{O}/\omega \\ (u, \omega) = 1}} \psi\left(\frac{1}{4m} u\omega^{-1} Na^2\right)$$

$$+ 2 \sum_{\substack{u \in \mathcal{O}/\omega \\ (u, \omega) = 1}} \sum_{\mu \in (\mathcal{O}/\omega)^{*2}} \psi\left(\frac{1}{4m} u\omega^{-1}(Na^2 + \mu)\right).$$

As Na^2 is a square modulo $\omega\mathcal{O}$, we have $Na^2 + \mu \in \omega\mathcal{O}$ for exactly one $\mu \in (\mathcal{O}/\omega)^{*2}$, so that

$$b(N, a) = (-1) + 2\left(\frac{q-1}{2} - 1\right)(-1) + 2(q - 1) = q.$$

Note that case 2 can only occur if q is odd, and that in the last case this assumption was used. Now if q is even, the map $\mu \mapsto \mu^2$ is the Frobenius automorphism of $\mathcal{O}/\omega$. Consequently one has

$$
\begin{aligned}
b(N, a) &= \sum_{\substack{u \in \mathcal{O}/\omega \\ (u, \omega) = 1}} \sum_{\mu \in \mathcal{O}/\omega} \psi\left(\frac{1}{4m} u\omega^{-1}(Na^2 + \mu)\right) \\
&= \sum_{\substack{u \in \mathcal{O}/\omega \\ (u, \omega) = 1}} \sum_{\mu \in \mathcal{O}/\omega} \psi\left(\frac{1}{4m} u\omega^{-1}\mu\right) \\
&= \sum_{\substack{u \in \mathcal{O}/\omega \\ (u, \omega) = 1}} \psi(0) + \sum_{\substack{u \in \mathcal{O}/\omega \\ (u, \omega) = 1}} \sum_{\substack{\mu \in \mathcal{O}/\omega \\ (\mu, \omega) = 1}} \psi\left(\frac{1}{4m} u\omega^{-1}\right) \\
&= (q - 1) + \sum_{\substack{\mu \in \mathcal{O}/\omega \\ (\mu, \omega) = 1}} (-1) = 0. \qquad \Box
\end{aligned}
$$

6.5 Local factors and the spherical dual

In this section we make our first attempt to attach local factors to irreducible, spherical representations of G^J. For some reductive groups such factors may be obtained as zeta integrals of spherical Whittaker functions. We have computed such functions in the previous section, hence we try this approach. The formula appearing in the following definition is a more or less natural generalization of the zeta integral from the GL(2)-theory, and is also inspired by the Mellin transform of a Jacobi form (see [Be5]), resp. an integral appearing in [Su] 4.

6.5.1 Definition. *Let $\mathcal{W}$ be a $\psi^{n,r}$-Whittaker model for the irreducible admissible representation π of G^J. Then for $W \in \mathcal{W}$, the* **zeta integral** *is defined as*

$$\zeta(W, s) = \int_{F^*} \int_F W(d(a, \lambda)) |a|^{s-3/2} \, d\lambda \, d^*a, \qquad s \in \mathbb{C}.$$

Notice the slight difference with the zeta integral defined in [Ho] 2.2.1. Using Proposition 5.7.3, it is not hard to prove that the integral above converges for $\mathrm{Re}(s) > s_0$, with s_0 independent of W, and represents a holomorphic function on this right half plane ([Ho] 2.2.2).

6.5.2 Proposition. *Let $v(2m) = 0$ and $\pi = \pi_{m,\chi}$ a spherical representation with Whittaker model $\mathcal{W}^{n,r}$. Let W be the spherical Whittaker function normalized as in Theorem 6.4.6.*

i) *If π is a principal series representation with $\chi^2 \neq |\,|, 1$, then*

$$\zeta(W, s) = q^{l(3/2-s)} \frac{b + cq^{-s}}{1 - (\chi(\omega) + \chi(\omega)^{-1})q^{-s} + q^{-2s}}$$

with

$$b = 1 - \beta \frac{\chi(\omega)^l - \chi(\omega)^{-l}}{\chi(\omega) - \chi(\omega)^{-1}}, \qquad c = \beta \frac{\chi(\omega)^{l-1} - \chi(\omega)^{1-l}}{\chi(\omega) - \chi(\omega)^{-1}}.$$

ii) *If π is a principal series representation with $\chi^2 = 1$, then*

$$\zeta(W, s) = q^{l(3/2-s)} \frac{b + cq^{-s}}{(1 - \chi(\omega)q^{-s})^2}$$

with

$$b = \chi(\omega)(1 - 2l\beta), \qquad c = \beta(2l - 1) + \chi(\omega) - 1.$$

iii) *If π is a positive Weil representation, then*

$$\zeta(W, s) = q^{-ls} \frac{\chi(\omega)^{-1}}{1 - \chi(\omega)^{-1}q^{-s}}.$$

Here we have as before $\beta = \left(\frac{-N\omega^{2l}}{\omega}\right) q^{-1/2}$, $N = 4mn - r^2$, l is defined in (6.15), and the symbol $\left(\frac{\cdot}{\omega}\right)$ in (6.17).

Proof: These are standard $\mathfrak{p}$-adic computations. We only go through one of them, namely when $\chi^2 \neq |\,|, 1$. According to Theorem 6.4.6,

$$W(d(a, \lambda)) = |a|^{3/2} 1_{\mathcal{O}}(ra + 2m\lambda)\Big(\chi(a)b + \chi(a)^{-1}c\Big)1_{\mathcal{O}}(Na^2)$$

with

$$b = q^{3l/2}\frac{\chi(\omega)^{1-l} - \beta}{\chi(\omega) - \chi(\omega)^{-1}}, \qquad c = q^{3l/2}\frac{\beta - \chi(\omega)^{l-1}}{\chi(\omega) - \chi(\omega)^{-1}}.$$

So we compute for $\mathrm{Re}(s)$ large enough

$$
\begin{aligned}
\zeta(W, s) &= \int_{F^*}\int_{F} 1_{\mathcal{O}}(ra + 2m\lambda)\Big(\chi(a)b + \chi(a)^{-1}c\Big)1_{\mathcal{O}}(Na^2)|a|^s \, d\lambda \, d^*a \\
&= \int_{F^*} \Big(\chi(a)b + \chi(a)^{-1}c\Big)1_{\mathcal{O}}(Na^2)|a|^s \, d^*a \\
&= \sum_{i \in \mathbb{Z}}\int_{\omega^i\mathcal{O}^*} \Big(\chi(a)b + \chi(a)^{-1}c\Big)1_{\mathcal{O}}(Na^2)|a|^s \, d^*a \\
&= \sum_{i \geq l}\int_{\omega^i\mathcal{O}^*} \Big(\chi(\omega)^i b + \chi(\omega)^{-i}c\Big)q^{-is} \, d^*a \\
&= b\sum_{i \geq l}\Big(\chi(\omega)q^{-s}\Big)^i + c\sum_{i \geq l}\Big(\chi(\omega)^{-1}q^{-s}\Big)^i \\
&= b\Big(\chi(\omega)q^{-s}\Big)^l\frac{1}{1 - \chi(\omega)q^{-s}} + c\Big(\chi(\omega)^{-1}q^{-s}\Big)^l\frac{1}{1 - \chi(\omega)^{-1}q^{-s}}.
\end{aligned}
$$

Inserting the values of b and c gives the desired result. The cases (ii) and (iii) are treated similarly; for (ii) one uses the formula

$$\sum_{i \geq l} i x^i = \frac{lx^l + (1-l)x^{l+1}}{(1-x)^2}, \qquad x \in \mathbb{C}, \ |x| < 1,$$

instead of the geometric series. $\qquad\qquad\square$

For the proof of this proposition it is not really necessary to have the spherical Whittaker function explicitly at hand. In fact, from the results of the previous section one easily obtains

$$\zeta(W, s) = \sum_{i \geq l} W_i q^{-is}.$$

Then the recursion formula in Lemma 6.4.4 and comparison of formal power series also gives the result.

Now we take the denominators of the fractions in this proposition as our local factors attached to irreducible, spherical representations.

6.5.3 Definition. Let $\pi_{\chi,m}$ be a spherical principal series representation of $G^J(F)$, where we still assume that $v(2m) = 0$. The **local Euler factor** $L(s,\pi)$ attached to π is defined as

$$L(s,\pi) := \frac{1}{(1 - \chi(\omega)q^{-s})(1 - \chi(\omega)^{-1}q^{-s})}.$$

6.5.4 Remark. We do not define local Euler factors for the remaining two spherical representations, which are positive Weil representations. The reason is that these representations do not appear as local components in global automorphic representations of the Jacobi group, as will follow later by the corresponding statement for the metaplectic group (Proposition 23 on p. 80 of [Wa1]) and Corollary 7.3.5.

Much of our discussion in the previous chapters can be summarized in the following commutative diagram, in which all the arrows are bijections.

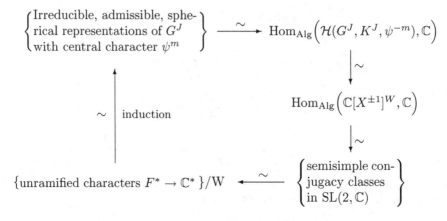

We explain the objects and maps in this diagram, starting in the lower left corner. Given an unramified character $\chi : F^* \to \mathbb{C}$, we associate to it the principal series representation $\pi_{\chi,m}$ if $\chi^2 \neq |\;|$, resp. the Weil representation $\sigma^+_{\xi,m}$ if $\chi = |\;|^{1/2}(\cdot,\xi)$ or $\chi = |\;|^{-1/2}(\cdot,\xi)$, with $v(\xi)$ even. If χ is replaced by χ^{-1}, then by Theorem 5.8.3 the same representation results. Hence if the nontrivial element in the Weyl group W of G^J operates on the unramified characters of F^* by $\chi \mapsto \chi^{-1}$, then we get the arrow indexed by 'induction'.

Given an irreducible, spherical representation with central character ψ^m, the Hecke algebra $\mathcal{H}(G^J, K^J, \psi^{-m})$ operates on the space of K^J-invariant vectors. This space is one-dimensional, and thus a character (algebra homomorphism) of $\mathcal{H}(G^J, K^J, \psi^{-m})$ is defined. This gives the upper horizontal map. If χ is the character we started with, then $\mathcal{H}(G^J, K^J, \psi^{-m}) \to \mathbb{C}$ is characterized by $T(\omega) \mapsto q^{3/2}(\chi(\omega) + \chi(\omega)^{-1})$, cf. Theorem 6.4.6. The upper arrow on the left is clear by the Satake isomorphism 6.2.8.

An algebra homomorphism $\mathbb{C}[X^{\pm 1}]^W \to \mathbb{C}$ clearly is determined by mapping X to a non-zero complex number z. Since we are dealing with polynomials which are invariant under $X \mapsto X^{-1}$, the complex numbers z and z^{-1} yield the same algebra homomorphism. As a result we can associate to the conjugacy class of $\begin{pmatrix} z & 0 \\ 0 & z^{-1} \end{pmatrix}$ in $\mathrm{SL}(2, \mathbb{C})$ the algebra homomorphism $\mathbb{C}[X^{\pm 1}]^W \to \mathbb{C}$ which maps $X + X^{-1}$ to $z + z^{-1}$, and every character of $\mathbb{C}[X^{\pm 1}]^W$ is thus obtained. This explains the lower left arrow. Finally, the lower horizontal arrow is induced by the map $\begin{pmatrix} z & 0 \\ 0 & z^{-1} \end{pmatrix} \mapsto \chi$, where $\chi(\omega) = z$.

6.5.5 Remark. The parametrization of spherical representations by semisimple conjugacy classes in the complex Lie group $\mathrm{SL}(2, \mathbb{C})$ offers another way to define local factors. As in the general reductive theory we could set

$$L(s, \pi) := \det\left(1 - gq^{-s}\right)^{-1},$$

where $g \in \mathrm{SL}(2, \mathbb{C})$ is any element in the conjugacy class corresponding to π. It is immediate from the above diagram that this factor coincides in the case of a principal series representation with the one defined in 6.5.3.

Now we can finish the proof of Theorem 6.3.10.

6.5.6 Corollary. *If F has odd residue characteristic and $v(m) = 0$, then super-cuspidal representations of G^J with central character ψ^m are not spherical.*

Proof: This is because there are simply no characters of the Hecke algebra left: They all come from induced representations. $\qquad\square$

6.6 The Eichler-Zagier operators

Let f be a classical Jacobi form of weight k and Index m, as defined in 4.1.1. There is a general lifting mechanism which assigns to f a function $\Phi = \Phi_f$ on the adelic Jacobi group $G^J(\mathbb{A})$, where $\mathbb{A}$ is the adele ring of the number field $\mathbb{Q}$. For the necessary global notions, see Section 7.1 below. The lifting procedure will be described in Section 7.4, but nevertheless, it seems apt to discuss the relationship between classical and adelic Hecke operators right now.

The lifted form Φ turns out to be right invariant under the local groups $G^J(\mathbb{Z}_p)$, for all finite places p. The local Hecke algebras $\mathcal{H}(G^J(\mathbb{Q}_p), G^J(\mathbb{Z}_p))$ operate on such global functions by convolution:

$$(\varphi.\Phi)(x) = \int_{G^J(\mathbb{Q}_p)} \varphi(y)\Phi(xy)\,dy, \qquad \varphi \in \mathcal{H}(G^J(\mathbb{Q}_p), G^J(\mathbb{Z}_p)).$$

If π denotes the representation by right translation, then this is just the corresponding representation of the Hecke algebra. But to fit in the context of classical Hecke operators, we let $\mathcal{H}(G^J(\mathbb{Q}_p), G^J(\mathbb{Z}_p))$ for the moment act on the right, and denote this action by $*$:

$$(\Phi * \varphi)(x) = \int_{G^J(\mathbb{Q}_p)} \varphi(y)\Phi(xy^{-1})\, dy, \qquad \varphi \in \mathcal{H}(G^J(\mathbb{Q}_p), G^J(\mathbb{Z}_p)).$$

It is not hard to see that if

$$\varphi = \sum_i \text{char}(G^J(\mathbb{Z}_p)g_i) \qquad \text{with } g_i \in G^J(\mathbb{Q}),$$

then the corresponding action on the Jacobi form f is given by

$$f\Big|_{k,m}\varphi = \sum_i f\Big|_{k,m}g_i.$$

The operator $|_{k,m}$ on the right-hand side is the one defined in [EZ], Theorem 1.4, or here in 4.1.1. In [EZ], §4, two more Hecke operators T_{EZ} und T_{EZ}^0 on classical Jacobi forms are defined:

$$f\Big|_{k,m}T_{EZ}(p^\alpha) :=$$

$$p^{\alpha(k-4)} \sum_{\substack{M \in \text{SL}_2(\mathbb{Z})\backslash\text{M}_2(\mathbb{Z}) \\ \det(M)=p^{2\alpha} \\ \gcd(M)=\square}} \sum_{X \in \mathbb{Z}^2/p^\alpha\mathbb{Z}^2} f\Big|_{k,m}\Big(\det(M)^{1/2}M(X,0)\Big),$$

$$f\Big|_{k,m}T_{EZ}^0(p^\alpha) :=$$

$$p^{\alpha(k-4)} \sum_{\substack{M \in \text{SL}_2(\mathbb{Z})\backslash\text{M}_2(\mathbb{Z}) \\ \det(M)=p^{2\alpha} \\ \gcd(M)=1}} \sum_{X \in \mathbb{Z}^2/p^\alpha\mathbb{Z}^2} f\Big|_{k,m}\Big(\det(M)^{1/2}M(X,0)\Big).$$

The condition $\gcd(M) = \square$ (resp. $= 1$) means summation over those matrices only where the greatest common divisor of all coefficients is a square number (resp. 1). These operators are now to be compared with the $T^J(p^\alpha)$. The following lemmas are valid for every number field, so we formulate them in greater generality than necessary.

6.6.1 Lemma. *Assume $\mathcal{O}$ to be the ring of integers of a $\mathfrak{p}$-adic field, and let $K^J = G^J(\mathcal{O})$. For $\gamma \in K^J \begin{pmatrix} \omega^{-\alpha} & 0 \\ 0 & \omega^\alpha \end{pmatrix} K^J$ let*

$$E_\gamma := K^J(\gamma \times H(\mathcal{O})) = \bigcup_{Y \in \mathcal{O}^2} K^J\gamma(Y,0).$$

Then:

$$\gamma \notin \mathrm{SL}(2,\mathcal{O})\begin{pmatrix} \omega^{-\alpha} & m\omega^{-\alpha} \\ 0 & \omega^{\alpha} \end{pmatrix}, \, m \in \mathcal{O} \implies E_\gamma = \coprod_{\lambda \in \mathcal{O}/\omega^\alpha} K^J \gamma(\lambda, 0, 0),$$

$$\gamma \in \mathrm{SL}(2,\mathcal{O})\begin{pmatrix} \omega^{-\alpha} & m\omega^{-\alpha} \\ 0 & \omega^{\alpha} \end{pmatrix}, \, m \in \mathcal{O} \implies E_\gamma = \coprod_{\mu \in \mathcal{O}/\omega^\alpha} K^J \gamma(0, \mu, 0).$$

Right multiplication by an element $g = (X, \kappa) \in H(\mathcal{O})$, $X \in \mathcal{O}^2$, $\kappa \in \mathcal{O}$, *induces a bijection* $E_\gamma \to E_\gamma$, *and permutes the cosets* $K^J \backslash E_\gamma$.

Proof: The coset decompositions are an exercise, and the other assertions are clear. □

6.6.2 Lemma. *With* γ *and* E_γ *as in the previous lemma, we have*

$$\sum_{\lambda, \mu \in \mathcal{O}/\omega^\alpha} \mathrm{char}\Big(K^J \gamma(\lambda, \mu, 0)\Big) = q^\alpha \mathrm{char}(E_\gamma).$$

Proof: Assume γ is not contained in the coset $\mathrm{SL}(2,\mathcal{O})\begin{pmatrix} \omega^{-\alpha} & m\omega^{-\alpha} \\ 0 & \omega^{\alpha} \end{pmatrix}$ (the other case is treated analogously). One computes

$$\sum_{\lambda, \mu \in \mathcal{O}/\omega^\alpha} \mathrm{char}\Big(K^J \gamma(\lambda, \mu, 0)\Big)$$

$$= \sum_{\mu \in \mathcal{O}/\omega^\alpha} \sum_{\lambda \in \mathcal{O}/\omega^\alpha} \mathrm{char}\Big(K^J \gamma(\lambda, 0, -\lambda\mu)(0, \mu, 0)\Big)$$

$$= \sum_{\mu \in \mathcal{O}/\omega^\alpha} \sum_{\lambda \in \mathcal{O}/\omega^\alpha} \mathrm{char}\Big(K^J \gamma(\lambda, 0, 0)(0, \mu, 0)\Big)$$

$$= \sum_{\mu \in \mathcal{O}/\omega^\alpha} \mathrm{char}(E_\gamma).$$

For the last step Lemma 6.6.1 was used. □

6.6.3 Lemma. *We have*

i) *the coset decomposition*

$$\{M \in \mathrm{M}_2(\mathcal{O}) : \det(M) = \omega^{2\alpha}\} = \coprod_{f=0}^{2\alpha} \coprod_{u \in \mathcal{O}/\omega^f} \mathrm{SL}(2,\mathcal{O})\begin{pmatrix} \omega^{2\alpha-f} & u \\ 0 & \omega^f \end{pmatrix},$$

ii) the decomposition

$$\{M \in M_2(\mathcal{O}) : \det(M) = \omega^{2\alpha},\ \gcd(M) = 1\}$$

$$= \coprod_{\substack{f=0}}^{2\alpha} \coprod_{\substack{u \in \mathcal{O}/\omega^f \\ (u,\omega^f,\omega^{2\alpha-f})=1}} \mathrm{SL}(2,\mathcal{O})\begin{pmatrix} \omega^{2\alpha-f} & u \\ 0 & \omega^f \end{pmatrix}$$

$$= \omega^\alpha\, \mathrm{SL}(2,\mathcal{O})\begin{pmatrix} \omega^{-\alpha} & 0 \\ 0 & \omega^\alpha \end{pmatrix}\mathrm{SL}(2,\mathcal{O}),$$

iii) and finally

$$\{M \in M_2(\mathbb{Z}) : \det(M) = p^{2\alpha},\ \gcd(M) = 1\}$$

$$= \coprod_{\substack{f=0}}^{2\alpha} \coprod_{\substack{u \in \mathbb{Z}/p^f \\ (u,p^f,p^{2\alpha-f})=1}} \mathrm{SL}(2,\mathcal{O})\begin{pmatrix} p^{2\alpha-f} & u \\ 0 & p^f \end{pmatrix}.$$

Proof: This is straightforward. $\square$

6.6.4 Lemma. *We have*

$$\{M \in M_2(\mathbb{Z}) : \det(M) = \omega^{2\alpha},\ \gcd(M) = \square\}$$

$$= \omega^\alpha \coprod_{j=0}^{\left[\frac{\alpha}{2}\right]} \coprod_{f=0}^{2(\alpha-2j)} \coprod_{\substack{u \in \mathcal{O}/\omega^f \\ (u,\omega^f,\omega^{2(\alpha-2j)-f})=1}} \mathrm{SL}(2,\mathcal{O})\begin{pmatrix} \omega^{(\alpha-2j)-f} & u\omega^{-(\alpha-2j)} \\ 0 & \omega^{f-(\alpha-2j)} \end{pmatrix}.$$

Proof: By Lemma 6.6.3 (i),

$$\{M \in M_2(\mathbb{Z}) : \det(M) = \omega^{2\alpha},\ \gcd(M) = \square\}$$

$$= \coprod_{j=0}^{\infty} \coprod_{f=0}^{2\alpha} \coprod_{\substack{u \in \mathcal{O}/\omega^f \\ (u,\omega^f,\omega^{2\alpha-f})=\omega^{2j}}} \mathrm{SL}(2,\mathcal{O})\begin{pmatrix} \omega^{2\alpha-f} & u \\ 0 & \omega^f \end{pmatrix}$$

$$= \coprod_{j=0}^{\left[\frac{\alpha}{2}\right]} \coprod_{f=2j}^{2\alpha-2j} \coprod_{\substack{u \in \mathcal{O}/\omega^f,\ \omega^{2j}|u \\ (u,\omega^f,\omega^{2\alpha-f})=\omega^{2j}}} \mathrm{SL}(2,\mathcal{O})\begin{pmatrix} \omega^{2\alpha-f} & u \\ 0 & \omega^f \end{pmatrix}$$

$$= \coprod_{j=0}^{\left[\frac{\alpha}{2}\right]} \coprod_{f=2j}^{2\alpha-2j} \coprod_{\substack{u \in \mathcal{O}/\omega^{f-2j} \\ (u,\omega^{f-2j},\omega^{2\alpha-f-2j})=1}} \mathrm{SL}(2,\mathcal{O}) \begin{pmatrix} \omega^{2\alpha-f} & u\omega^{2j} \\ 0 & \omega^f \end{pmatrix}$$

$$= \coprod_{j=0}^{\left[\frac{\alpha}{2}\right]} \coprod_{f=0}^{2(\alpha-2j)} \coprod_{\substack{u \in \mathcal{O}/\omega^f \\ (u,\omega^f,\omega^{2(\alpha-2j)-f})=1}} \mathrm{SL}(2,\mathcal{O}) \begin{pmatrix} \omega^{2\alpha-f-2j} & u\omega^{2j} \\ 0 & \omega^{f+2j} \end{pmatrix} \qquad \square$$

In the following proposition, which makes the connection between T^J and the Eichler-Zagier operators, we return to $F = \mathbb{Q}$, $\mathfrak{p} = (p)$, $\omega = p$.

6.6.5 Proposition.

i) $\qquad T_{EZ}^0(p^\alpha) = p^{k+\alpha-4} T^J(p^\alpha).$

ii) $\qquad T_{EZ}(p^\alpha) = \sum_{j=0}^{\left[\frac{\alpha}{2}\right]} p^{2j(k-2)} T_{EZ}^0(p^{\alpha-2j}).$

Proof: (i) It is not hard to see that if γ runs over a complete set of representatives of $G(\mathbb{Z}_p) \backslash G(\mathbb{Z}_p) \begin{pmatrix} p^\alpha & 0 \\ 0 & p^{-\alpha} \end{pmatrix} G(\mathbb{Z}_p)$, $G = \mathrm{SL}(2)$, and for every γ the element g runs over a set of representatives of $G^J(\mathbb{Z}_p) \backslash (\gamma \times H(\mathbb{Z}_p))$, then g runs over a set of representatives of $G^J(\mathbb{Z}_p) \backslash G^J(\mathbb{Z}_p) \begin{pmatrix} p^\alpha & 0 \\ 0 & p^{-\alpha} \end{pmatrix} G^J(\mathbb{Z}_p)$. Hence we can compute

$$f\Big|_{k,m} T^J(p^\alpha) = \sum_\gamma \sum_{g \in G^J(\mathbb{Z}_p) \backslash E_\gamma} f\big| g$$

$$\overset{\text{Lemma 6.6.2}}{=} \sum_\gamma \sum_{X \in \mathbb{Z}^2/p^\alpha \mathbb{Z}^2} \frac{1}{p^\alpha} f\big|(\gamma(X,0))$$

$$\overset{\text{Lemma 6.6.3}}{=} \frac{1}{p^\alpha} \sum_{\substack{M \in \mathrm{SL}_2(\mathbb{Z}) \backslash M_2(\mathbb{Z}) \\ \det(M)=p^{2\alpha} \\ \gcd(M)=1}} \sum_{X \in \mathbb{Z}^2/p^\alpha \mathbb{Z}^2} f\big|\big(\det(M)^{1/2} M(X,0)\big)$$

$$= \quad p^{-\alpha} p^{4-k} f\big| T_{EZ}^0(p^\alpha).$$

(ii) With the help of Lemma 6.6.4 one computes

$$
f\Big|T_{EZ}(p^\alpha) = p^{\alpha(k-4)} \sum_{j=0}^{\left[\frac{\alpha}{2}\right]} \sum_{f=0}^{2(\alpha-2j)} \sum_{\substack{u\in\mathcal{O}/\omega^f \\ (u,\omega^f,\omega^{2(\alpha-2j)-f}=1)}} \sum_{X\in\mathbb{Z}^2/p^{\alpha-2j}\mathbb{Z}^2}
$$

$$
\sum_{Y\in\mathbb{Z}^2/p^{2j}\mathbb{Z}^2} f\left|\left(\begin{pmatrix} p^{(\alpha-2j)-f} & up^{-(\alpha-2j)} \\ 0 & p^{f-(\alpha-2j)} \end{pmatrix}(X+p^{\alpha-2j}Y,0)\right)\right.
$$

$$
= \ p^{\alpha(k-4)} \sum_{j=0}^{\left[\frac{\alpha}{2}\right]} \sum_{\substack{M\in SL_2(\mathbb{Z})\backslash M_2(\mathbb{Z}) \\ \det(M)=p^{2(\alpha-2j)} \\ \gcd(M)=1}} p^{4j} \sum_{X\in\mathbb{Z}^2/p^{\alpha-2j}\mathbb{Z}^2} f\left|\left(p^{-(\alpha-2j)}M,X,0\right)\right.
$$

$$
= \ p^{\alpha(k-4)} \sum_{j=0}^{\left[\frac{\alpha}{2}\right]} p^{4j} p^{(\alpha-2j)(4-k)} \left(f\Big|T^0_{EZ}(p^{\alpha-2j})\right). \qquad\qquad \square
$$

7

Global Considerations

After having classified all unitary, resp. admissible, representations of $G^J(F)$, where F is a local field of characteristic zero, in the preceding chapters, we are now ready to consider representations of $G^J(\mathbb{A})$, where $\mathbb{A}$ is the adele ring of some number field. The first section of this chapter collects some basic results about the adelized Jacobi group.

After that, we consider once more the Schrödinger-Weil representation, this time in the global context. It will be shown how an *automorphic* version of π_{SW}^m is constructed by means of theta functions.

One of the main results will be an explicit version in the global context of the fundamental relation

$$\pi = \tilde{\pi} \otimes \pi_{SW}^m,$$

yielding a *canonical* bijection between automorphic representations of G^J (with fixed non-trivial central character) and genuine automorphic representations of the metaplectic group (see Theorem 7.3.3).

Similar to the situation for some reductive groups, the first examples of automorphic representations of G^J come from classical Jacobi cusp forms on $\mathbf{H} \times \mathbb{C}$. These can be lifted to the adelized group $G^J(\mathbb{A})$ ($\mathbb{A}$ here the adeles of $\mathbb{Q}$), thereby yielding an element of a certain cuspidal L^2-space. We describe this lifting procedure in Section 7.4.

After that, one has to prove that the subrepresentation of the right regular representation generated by this lifted function is irreducible, provided we start with an eigenform. This will be achieved by the help of a strong multiplicity-one result for the metaplectic group by Waldspurger and Gelbart, Piatetski-Shapiro, carried over to the Jacobi group by means of the above mentioned explicit isomorphism.

7.1 Adelization of G^J

In this chapter we use the following notations.

- F denotes a number field,

- $\mathcal{O}$ is its ring of integers,

- $\{\mathfrak{p}\}$ is the set of all places of F,

- $F_{\mathfrak{p}}$ is the completion of F at $\mathfrak{p}$,

- $\mathcal{O}_{\mathfrak{p}}$ is the closure of $\mathcal{O}$ in $F_{\mathfrak{p}}$,

- $\mathbb{A}$ is the ring of adeles of F.

The character ψ of $\mathbb{A}$ is defined to be the product of the local additive standard characters we used before, as explained in Section 2.2. It is a character of $F\backslash\mathbb{A}$, i.e.

$$\psi(\varrho + x) = \psi(x) \qquad \text{for all } x \in \mathbb{A}, \ \varrho \in F.$$

For an adele $m \in \mathbb{A}$ the symbol ψ^m denotes the character $x \mapsto \psi(mx)$ of $\mathbb{A}$. The characters of $F\backslash\mathbb{A}$ are then exactly the ψ^m with $m \in F$.

The adelization of the Jacobi group is defined as

$$G^J(\mathbb{A}) = \prod_{\mathfrak{p}} \Big(G^J(F_{\mathfrak{p}}) : G^J(\mathcal{O}_{\mathfrak{p}})\Big),$$

the restricted direct product of the local Jacobi groups $G^J(F_{\mathfrak{p}})$ with respect to the open compact subgroups $G^J(\mathcal{O}_{\mathfrak{p}})$ at the finite places.

The group $G^J(F)$ is embedded diagonally in $G^J(\mathbb{A})$ as a discrete subgroup. This follows from the analogous statements for the groups $G = \mathrm{SL}(2)$ and H. We will be concerned with the homogeneous space

$$G^J(F)\backslash G^J(\mathbb{A}).$$

One checks that the following integration formula for suitable functions Φ on $G^J(F)\backslash G^J(\mathbb{A})$ holds:

$$\int\limits_{G^J(F)\backslash G^J(\mathbb{A})} \Phi(g)\, dg = \int\limits_{G(F)\backslash G(\mathbb{A})} \int\limits_{H(F)\backslash H(\mathbb{A})} \Phi(hM)\, dh\, dM.$$

($G = \mathrm{SL}(2)$ as before). Since the measure of $G(F)\backslash G(\mathbb{A})$ is finite (well known) and that of $H(F)\backslash H(\mathbb{A})$ is too (being compact), it follows from this formula that

$$G^J(F)\backslash G^J(\mathbb{A}) \text{ has finite measure.} \tag{7.1}$$

For $G(\mathbb{A})$ we have the well known strong approximation theorem, whereafter $G(F)G_\infty$ is dense in $G(\mathbb{A})$, the ∞ always denoting all the infinite components. Strong approximation does also hold for the Heisenberg group, because $F\mathbb{A}_\infty$ is dense in $\mathbb{A}$. Thus we conclude that the Jacobi group fulfills strong approximation, too:

$$G^J(F)G^J_\infty \text{ is dense in } G^J(\mathbb{A}). \tag{7.2}$$

In particular, with

$$K^J_0 := \prod_{\mathfrak{p}\nmid\infty} G^J(\mathcal{O}_\mathfrak{p})$$

it follows that

$$G^J(\mathbb{A}) = G^J(F)G^J_\infty K^J_0. \tag{7.3}$$

We look at the special case $F = \mathbb{Q}$. By (7.3), the injection $G^J(\mathbb{R}) \hookrightarrow G^J(\mathbb{A})$ yields a bijection

$$\Gamma^J\backslash G^J(\mathbb{R})/K_\infty Z(\mathbb{R}) \simeq G^J(\mathbb{Q})\backslash G^J(\mathbb{A})/K_\infty K_0 Z(\mathbb{A}), \tag{7.4}$$

which is easily seen to be a homeomorphism. Here we remind the reader that we put $K_\infty = \mathrm{SO}(2,\mathbb{R})$, and Z is the center of G^J. As already remarked in Section 1.4, the real Jacobi group $G^J(\mathbb{R})$ acts on $\mathbf{H} \times \mathbb{C}$ in the following way:

$$\begin{pmatrix} a & b \\ c & d \end{pmatrix}(\lambda,\mu,\kappa).(\tau,z) = \left(\frac{a\tau + b}{c\tau + d}, \frac{z + \lambda\tau + \mu}{c\tau + d} \right).$$

In particular,

$$hM(i,0) = (\tau, \lambda\tau + \mu), \qquad \text{where } \tau = M(i), \ h = (\lambda,\mu,\kappa).$$

The stabilizer of the special point $(i,0) \in \mathbf{H} \times \mathbb{C}$ is the group $K_\infty Z(\mathbb{R})$. Hence there is a homeomorphism

$$G^J(\mathbb{R})/K_\infty Z(\mathbb{R}) \simeq \mathbf{H} \times \mathbb{C}. \tag{7.5}$$

Taking (7.4) and (7.5) together, we see that there is a homeomorphism

$$\Gamma^J\backslash \mathbf{H} \times \mathbb{C} \simeq G^J(\mathbb{Q})\backslash G^J(\mathbb{A})/K_\infty K_0 Z(\mathbb{A}). \tag{7.6}$$

Classical Jacobi forms may therefore be lifted to functions on the homogeneous space $G^J(\mathbb{Q})\backslash G^J(\mathbb{A})$. This will be carried out in Section 7.4. But before that, we discuss the Schrödinger-Weil representation, which is fundamental for the representation theory of G^J in the global case also.

7.2   The global Schrödinger-Weil representation

First we introduce the **global Schrödinger representation** π_S^m of the adelized Heisenberg group $H(\mathbb{A})$ on the Hilbert space $L^2(\mathbb{A})$. It is the unique unitary representation which acts on the dense subspace $\mathcal{S}(\mathbb{A})$ of $L^2(\mathbb{A})$ as

$$\left(\pi_S^m(\lambda, \mu, \kappa)f\right)(x) = \psi^m(\kappa + (2x + \lambda)\mu)f(x + \lambda) \qquad (\lambda, \mu, \kappa \in \mathbb{A}).$$

It is easily seen that the global Schrödinger representation is the tensor product of local ones. Hence from the local Stone–von Neumann theorems one deduces:

7.2.1 Theorem. (The global Stone–von Neumann theorem)

    i) π_S^m *is the unique irreducible unitary representation of $H(\mathbb{A})$ with central character ψ^m.*

    ii) *Every smooth unitary representation of $H(\mathbb{A})$ with central character ψ^m is isomorphic to a direct sum of Schrödinger representations π_S^m.*

Just as in the local case, this theorem allows us to construct a **global Weil representation**, which will be a projective representation of the adelized group $G(\mathbb{A})$, or alternatively, a representation of a certain two-fold cover of $G(\mathbb{A})$, namely the **global metaplectic group** $\mathrm{Mp}(\mathbb{A})$. We collect some facts about this group. To begin with, there is a short exact sequence of topological groups

$$1 \longrightarrow \{\pm 1\} \longrightarrow \mathrm{Mp}(\mathbb{A}) \longrightarrow G(\mathbb{A}) \longrightarrow 1. \tag{7.7}$$

This sequence does not split, but is given by a nontrivial cocycle

$$\lambda \in H^2(G(\mathbb{A}), \{\pm 1\}).$$

Hence $\mathrm{Mp}(\mathbb{A})$ may be realized as the set $G(\mathbb{A}) \times \{\pm 1\}$, endowed with the multiplication

$$(M, \varepsilon)(M', \varepsilon') = (MM', \lambda(M, M')\varepsilon\varepsilon')$$

for all $M, M' \in G(\mathbb{A})$, $\varepsilon, \varepsilon' \in \{\pm 1\}$. The cocycle λ is the product of local cocycles $\lambda_{\mathfrak{p}}$ for all places $\mathfrak{p}$ of F:

$$\lambda(M, M') = \prod_{\mathfrak{p}} \lambda_{\mathfrak{p}}(M_{\mathfrak{p}}, M'_{\mathfrak{p}}) \qquad \text{for } M = (M_{\mathfrak{p}}),\ M' = (M'_{\mathfrak{p}}).$$

This is well defined because the $\lambda_{\mathfrak{p}}$ were designed in such a way that for $\mathfrak{p} \nmid 2$ the cocycle $\lambda_{\mathfrak{p}}$ is trivial on $G(\mathcal{O}_{\mathfrak{p}}) \times G(\mathcal{O}_{\mathfrak{p}})$. Unfortunately, a product formula does not hold for the $\lambda_{\mathfrak{p}}$ in the sense that $\lambda(M, M') = 1$ for all $M, M' \in G(F)$ would hold. However, there is something which is almost as good (cf. [Ge2] 2.2 or [Sch1] 5.2):

7.2.2 Lemma. *There is a function* $\eta : G(\mathbb{A}) \to \{\pm 1\}$ *such that*

$$\lambda(M, M') = \eta(M)\eta(M')\eta(MM') \qquad \text{for all } M, M' \in G(F).$$

Therefore the sequence (7.7) splits over $G(F)$, and we have an injection

$$
\begin{aligned}
G(F) &\longrightarrow \text{Mp}(\mathbb{A}), \\
M &\longmapsto (M, \eta(M)).
\end{aligned}
$$

The image of this map is denoted by $\text{Mp}(F)$. This is something like the F-rational points of Mp, although the metaplectic group is not algebraic and $\text{Mp}(\mathbb{A})$ is not quite the restricted direct product of the local groups $\text{Mp}(F_{\mathfrak{p}})$.

We turn back to the global Weil representation π_W^m of $G(\mathbb{A})$ resp. $\text{Mp}(\mathbb{A})$ on $L^2(\mathbb{A})$ and give some explicit formulas for it. These look slightly simpler than in the local cases, due to the fact that there is a product formula for the Weil constant γ_m. For every $\Phi \in \mathcal{S}(\mathbb{A})$, $b, x \in \mathbb{A}$, and $a \in I$ (the idel group of F) the following holds:

$$
\begin{aligned}
\left(\pi_W^m \begin{pmatrix} 1 & b \\ 0 & 1 \end{pmatrix} \Phi \right)(x) &= \psi^m(bx^2)\Phi(x), \\
\left(\pi_W^m \begin{pmatrix} a & 0 \\ 0 & a^{-1} \end{pmatrix} \Phi \right)(x) &= |a|^{1/2}\Phi(ax), \\
(\pi_W^m(w)\Phi)(x) &= \hat{\Phi}(x) = \int_{\mathbb{A}} \Phi(y)\psi^m(2xy)\, dy.
\end{aligned}
$$

Here w denotes as usual the element $\begin{pmatrix} 0 & 1 \\ -1 & 0 \end{pmatrix}$ of $G(F) \subset G(\mathbb{A})$, and $|\ |$ is the product over all places of the absolute values of $F_{\mathfrak{p}}$, normalized such that the product formula holds. As in the local cases, the Schrödinger and the Weil representation can be put together to obtain the **global Schrödinger-Weil representation** π_{SW}^m of $G^J(\mathbb{A})$ (resp. of a two-fold cover $\widetilde{G}^J(\mathbb{A})$).

For the next section it is important to have an automorphic version of the Schrödinger representation. From now on the element $m \in \mathbb{A}$ is chosen from F. To every function $f \in \mathcal{S}(\mathbb{A})$ we associate a **theta function** $\vartheta_f : H(\mathbb{A}) \to \mathbb{C}$ by

$$\vartheta_f(h) = \sum_{\xi \in F} (\pi_S^m(h)f)(\xi), \qquad h \in H(\mathbb{A}).$$

It is not hard to see that this series converges and defines a continuous function on $H(\mathbb{A})$ which is right invariant by $H(F)$. Furthermore it satisfies the relation

$\vartheta(hz) = \psi^m(z)\vartheta(h)$ for all $h \in H(\mathbb{A})$ and $z \in Z(\mathbb{A})$. We compute its L^2-norm:

$$\|\vartheta_f\|_2^2 = \int\limits_{H(F)\backslash H(\mathbb{A})/Z(\mathbb{A})} |\vartheta_f(h)|^2 \, dh$$

$$= \int\limits_{H(F)\backslash H(\mathbb{A})/Z(\mathbb{A})} \left| \sum_\xi (\pi_s^m(h)f)(\xi) \right|^2 dh$$

$$= \int\limits_{F\backslash\mathbb{A}} \int\limits_{F\backslash\mathbb{A}} \left| \sum_\xi \psi^m((2\xi+\lambda)\mu)f(\xi+\lambda) \right|^2 d\mu \, d\lambda$$

$$= \int\limits_{F\backslash\mathbb{A}} \int\limits_{F\backslash\mathbb{A}} \sum_{\xi_1,\xi_2} \psi^m((2\xi_1+\lambda)\mu - (2\xi_2+\lambda)\mu)f(\xi_1+\lambda)\overline{f(\xi_2+\lambda)} \, d\mu \, d\lambda$$

$$= \int\limits_{F\backslash\mathbb{A}} \sum_\xi |f(\xi+\lambda)|^2 \, d\lambda$$

$$= \int\limits_{\mathbb{A}} |f(\lambda)|^2 \, d\lambda = \|f\|_2^2. \tag{7.8}$$

The map $f \mapsto \vartheta_f$ may therefore be extended to a norm-preserving map of Hilbert spaces

$$\vartheta : L^2(\mathbb{A}) \longrightarrow L^2(H(F)\backslash H(\mathbb{A}))_m,$$

where the space on the right consists of all measurable functions $\Phi : H(\mathbb{A}) \to \mathbb{C}$ which fulfill

$$\Phi(\varrho h z) = \psi^m(z)\Phi(h) \qquad \text{for all } \varrho \in H(F), \ h \in H(\mathbb{A}), \ z \in Z(\mathbb{A}),\tag{7.9}$$

and

$$\int\limits_{H(F)\backslash H(\mathbb{A})/Z(\mathbb{A})} |\Phi(h)|^2 \, dh < \infty.$$

If we associate to such a Φ the function

$$f_\Phi : \mathbb{A} \longrightarrow \mathbb{C},$$

$$\lambda \longmapsto \int\limits_{F\backslash\mathbb{A}} \Phi((0,\mu,0)(\lambda,0,0)) \, d\mu,$$

then, using some standard Fourier analysis on $F\backslash\mathbb{A}$, it is not hard to see that $\Phi \mapsto f_\Phi$ gives an inverse map to ϑ. In particular, for $f \in \mathcal{S}(\mathbb{A})$,

$$f(\lambda) = \int\limits_{F\backslash\mathbb{A}} \vartheta_f((0,\mu,0)(\lambda,0,0)) \, d\mu \qquad \text{for all } \lambda \in \mathbb{A}.\tag{7.10}$$

Thus ϑ is an isomorphism of Hilbert spaces which intertwines the $H(\mathbb{A})$-action on both sides. In particular, the right regular representation on the space $L^2(H(F)\backslash H(\mathbb{A}))_m$ is irreducible, which may be interpreted as a multiplicity-one result for automorphic representations of the Heisenberg group. If we consider $H(\mathbb{A})$ as a subgroup of $G^J(\mathbb{A})$ and conjugate $H(F)$ by an element of $G(\mathbb{A})$, we obtain the following generalization.

7.2.3 Proposition. *For any $M \in G(\mathbb{A})$, denote by $H(F)^M$ the conjugate group $M^{-1}H(F)M$. Then the right regular representation of $H(\mathbb{A})$ on the space $L^2(H(F)^M \backslash H(\mathbb{A}))_m$ is isomorphic to π_S^m. In particular, it is irreducible.*

It is also possible to lift the Schrödinger-Weil representation from $L^2(\mathbb{A})$ to functions living on the Jacobi group. Just assign to $f \in \mathcal{S}(\mathbb{A})$ the theta function

$$\vartheta_f(g) = \sum_{\xi \in F} (\pi_{SW}^m(g)f)(\xi), \qquad g \in G(\mathbb{A}).$$

This is an isometrical map if the norm on the image is given by integration over the Heisenberg group only, just as in (7.9). If the image of $L^2(\mathbb{A})$ is denoted by $\mathcal{R}_{SW}^m$, then we can state the following.

7.2.4 Proposition. *The representation of $G^J(\mathbb{A})$ on $\mathcal{R}_{SW}^m$ by right translation is isomorphic to π_{SW}^m. For every element $\vartheta \in \mathcal{R}_{SW}^m$ the following holds:*

$$\vartheta(\varrho g) = \eta(\varrho)\lambda(\varrho, g)\vartheta(g) \qquad \text{for all } \varrho \in G^J(F), \, g \in G^J(\mathbb{A}).$$

Here the functions η and λ are carried over trivially from $G(\mathbb{A})$ to $G^J(\mathbb{A})$.

We do not give the proof of this invariance property, which rests on the Poisson summation formula (cf. [We]).

7.3 Automorphic representations

Let $L^2(G^J(F)\backslash G^J(\mathbb{A}))_m$ be the Hilbert space of measurable functions $\Phi : G^J(\mathbb{A}) \to \mathbb{C}$ which satisfy

$$\Phi(\varrho g z) = \psi^m(z)\Phi(g) \qquad \text{for all } g \in G^J(\mathbb{A}), \, \varrho \in G^J(F), \, z \in Z(\mathbb{A})$$

and

$$\int\limits_{G^J(F)\backslash G^J(\mathbb{A})/Z(\mathbb{A})} |\Phi(g)|^2 \, dg < \infty.$$

Similarly, let $L^2(\mathrm{Mp}(F)\backslash \mathrm{Mp}(\mathbb{A}))_-$ be the Hilbert space of measurable functions $\varphi : \mathrm{Mp}(\mathbb{A}) \to \mathbb{C}$ which satisfy

$$\varphi(\varrho(g, \varepsilon)) = \varepsilon\varphi(g, 1) \qquad \text{for all } \varrho \in \mathrm{Mp}(F), \, g \in G(\mathbb{A}), \, \varepsilon \in \{\pm 1\}$$

and

$$\int\limits_{\mathrm{Mp}(F)\backslash\mathrm{Mp}(\mathbb{A})/S} |\varphi(g)|^2\, dg = \int\limits_{G(F)\backslash G(\mathbb{A})} |\varphi(M,1)|^2\, dM < \infty,$$

where we have denoted by S the two-element subgroup $(\mathbf{1}, \pm 1) \in \mathrm{Mp}(\mathbb{A})$.

7.3.1 Definition. *An irreducible, unitary representation of $G^J(\mathbb{A})$ (resp. of $\mathrm{Mp}(\mathbb{A})$) is called* **automorphic** *if it appears as a subrepresentation of the right regular representation of $G^J(\mathbb{A})$ (resp. $\mathrm{Mp}(\mathbb{A})$) on $L^2(G^J(F)\backslash G^J(\mathbb{A}))_m$ (resp. $L^2(\mathrm{Mp}(F)\backslash\mathrm{Mp}(\mathbb{A}))_-$).*

The following lemma is decisive.

7.3.2 Lemma. *Let $U \subset L^2(G^J(F)\backslash G^J(\mathbb{A}))_m$ be a subspace which is invariant under right translation by elements of $H(\mathbb{A})$, and such that the representation of $H(\mathbb{A})$ thus defined is equivalent to π_s^m. Then there exists a function φ in $L^2(\mathrm{Mp}(F)\backslash\mathrm{Mp}(\mathbb{A}))_-$ such that every $\Phi \in U$ is of the form*

$$\Phi(Mh) = \varphi(M,1)\vartheta_f(Mh) \qquad \text{for all } M \in G(\mathbb{A}),\ h \in H(\mathbb{A}),$$

for some $f \in L^2(\mathbb{A})$.

Proof: For $M \in G(\mathbb{A})$ fixed, we compare the map

$$\begin{aligned} U &\xrightarrow{\sim} L^2(H(F)^M\backslash H(\mathbb{A})), \\ \Phi &\longmapsto \Big(h \mapsto \Phi(Mh)\Big), \end{aligned}$$

which intertwines the $H(\mathbb{A})$-action and is an isomorphism by Proposition 7.2.3, to the composition of $H(\mathbb{A})$-intertwining maps

$$U \xrightarrow{\sim} L^2(H(F)\backslash H(\mathbb{A}))_m \xrightarrow{\sim} \mathcal{R}_{sw}^m \xrightarrow{\sim} L^2(H(F)^M\backslash H(\mathbb{A}))_m, \tag{7.11}$$

where the first map is restriction, the inverse of the second is also restriction, and the third one maps ϑ to the function $h \mapsto \vartheta(Mh)$ (the space $\mathcal{R}_{sw}^m$ is defined before Proposition 7.2.4). By Schur's lemma, the two isomorphisms $U \to L^2(H(F)^M\backslash H(\mathbb{A}))$ thus obtained differ by a scalar, which we denote by $\varphi(M)$. This defines the function φ on $G(\mathbb{A})$. Going through the definitions, we see that

$$\Phi(Mh) = \varphi(M)\vartheta_f(Mh) \qquad \text{for all } M \in G(\mathbb{A}),\ h \in H(\mathbb{A}),$$

where ϑ_f is the image of Φ under the first map in (7.11). Now, using (7.8), it is easy to see that φ satisfies the correct L^2-condition, which allows us to lift it to a function in $L^2(\mathrm{Mp}(F), \mathrm{Mp}(\mathbb{A}))_-$. $\qquad\square$

Now we can state the main result of this section, which makes the abstract isomorphism $\pi \simeq \tilde{\pi} \otimes \pi_{SW}^m$ from Section 2.6 concrete in the global case. Similar statements can be found in the real case in Takase [Ta] (Section 11) and in the adelic case in Pyatetski-Shapiro [PS2] (equation 6.8).

7.3.3 Theorem. *There is a natural isomorphism of Hilbert spaces*

$$L^2(\mathrm{Mp}(F)\backslash\mathrm{Mp}(\mathbb{A}))_- \otimes L^2(\mathbb{A}) \;\xrightarrow{\sim}\; L^2(G^J(F)\backslash G^J(\mathbb{A}))_m,$$

$$\varphi \otimes f \;\longmapsto\; \Big(Mh \mapsto \varphi(M,1)\vartheta_f(Mh)\Big).$$

This isomorphism is an intertwining map for the representation of $G^J(\mathbb{A})$ on both sides, where the action of $G^J(\mathbb{A})$ on $L^2(\mathbb{A})$ is the Schrödinger-Weil representation π_{SW}^m, and right translation on the other parts. Restriction to cuspidal functions yields another isomorphism

$$L_0^2(\mathrm{Mp}(F)\backslash\mathrm{Mp}(\mathbb{A}))_- \otimes L^2(\mathbb{A}) \;\xrightarrow{\sim}\; L_0^2(G^J(F)\backslash G^J(\mathbb{A}))_m.$$

Proof: Denoting by Φ the image of $\varphi \otimes f$, it is easy to see by Proposition 7.2.4 that Φ has the correct transformation property. We use the calculation (7.8) to see that Φ is indeed an L^2-function and the map is norm-preserving:

$$\int\limits_{G^J(F)\backslash G^J(\mathbb{A})/Z(\mathbb{A})} |\Phi(g)|^2 \, dg = \int\limits_{G(F)\backslash G(\mathbb{A})} \int\limits_{H(F)\backslash H(\mathbb{A})/Z(\mathbb{A})} |\Phi(hM)|^2 \, dh \, dM$$

$$= \int\limits_{G(F)\backslash G(\mathbb{A})} |\varphi(M,1)|^2 \, \|\pi_W^m(M)f\|^2 \, dm$$

$$= \int\limits_{G(F)\backslash G(\mathbb{A})} |\varphi(M,1)|^2 \, \|f\|^2 \, dm = \|\varphi\|^2 \, \|f\|^2.$$

By the preceding lemma and the global Stone–von Neumann theorem, our map is surjective, whence the first isomorphism. Now suppose U is a subspace like in Lemma 7.3.2 which lies in the cuspidal subspace $L_0^2(G^J(F)\backslash G^J(\mathbb{A}))_m$. If φ is the function appearing in this lemma, the element f in $\mathcal{S}(\mathbb{A})$ is arbitrary, and $\Phi = \varphi \otimes \vartheta_f \in U$, then for almost all $g = Mh \in G^J(\mathbb{A})$

$$0 = \int\limits_{N^J(F)\backslash N^J(\mathbb{A})} \Phi(ng) \, dn$$

$$= \int\limits_{F\backslash\mathbb{A}} \int\limits_{F\backslash\mathbb{A}} \varphi\left(\begin{pmatrix} 1 & x \\ 0 & 1 \end{pmatrix}M, 1\right) \vartheta_f\left(\begin{pmatrix} 1 & x \\ 0 & 1 \end{pmatrix}(0,\mu,0)Mh\right) \, dx \, d\mu$$

$$\overset{(7.10)}{=} \int\limits_{F\backslash\mathbb{A}} \varphi\left(\begin{pmatrix} 1 & x \\ 0 & 1 \end{pmatrix}M, 1\right) \left(\pi_{SW}\left(\begin{pmatrix} 1 & x \\ 0 & 1 \end{pmatrix}Mh\right)f\right)(0) \, dx$$

$$= \int_{F\backslash \mathbb{A}} \varphi\left(\begin{pmatrix} 1 & x \\ 0 & 1 \end{pmatrix} M, 1\right) dx \cdot \left(\pi_{SW}(Mh)f\right)(0).$$

It is easy to see that h can be chosen such that $(\pi_{SW}(Mh)f)(0) \neq 0$. It follows that

$$\int_{F\backslash \mathbb{A}} \varphi\left(\begin{pmatrix} 1 & x \\ 0 & 1 \end{pmatrix} M, 1\right) dx = 0 \qquad \text{for almost all } M \in G(\mathbb{A}),$$

i.e., φ is cuspidal. □

7.3.4 Corollary. *Every automorphic representation π of G^J factorizes as a restricted tensor product*

$$\pi = \bigotimes_{\mathfrak{p}}' \pi_{\mathfrak{p}}$$

with irreducible, unitary representations of the local groups $G^J(F_{\mathfrak{p}})$.

Proof: The corresponding statement for automorphic representations $\tilde{\pi}$ of the metaplectic group is known to be true, cf. [F]. For suitable $\tilde{\pi}$ we therefore have

$$\pi \simeq \tilde{\pi} \otimes \pi_{SW}^m \quad \simeq \quad \left(\bigotimes_{\mathfrak{p}}' \tilde{\pi}_{\mathfrak{p}}\right) \otimes \left(\bigotimes_{\mathfrak{p}}' \pi_{SW,\mathfrak{p}}^m\right) \simeq \bigotimes_{\mathfrak{p}}' \left(\tilde{\pi}_{\mathfrak{p}} \otimes \pi_{SW,\mathfrak{p}}^m\right). \quad □$$

7.3.5 Corollary. *The map*

$$\tilde{\pi} \longmapsto \pi := \tilde{\pi} \otimes \pi_{SW}^m$$

gives a 1-1 correspondence between (genuine) automorphic representations of $\mathrm{Mp}(\mathbb{A})$ and automorphic representations of $G^J(\mathbb{A})$ with central character ψ^m. The cuspidal representations correspond to the cuspidal ones. Further, this correspondence is compatible with local data: If

$$\tilde{\pi} = \bigotimes_{\mathfrak{p}}' \tilde{\pi}_{\mathfrak{p}}, \qquad \pi = \bigotimes_{\mathfrak{p}}' \pi_{\mathfrak{p}}$$

are the decompositions of $\tilde{\pi}$ and π in local components, then

$$\pi_{\mathfrak{p}} = \tilde{\pi}_{\mathfrak{p}} \otimes \pi_{SW,\mathfrak{p}}^m.$$

7.3.6 Corollary. *The space $L_0^2(G^J(F)\backslash G^J(\mathbb{A}))_m$ decomposes in a discrete direct sum of irreducible representations, each occurring with finite multiplicity.*

Proof: This is true for the metaplectic group. The proof runs along the lines sketched in [Ge1], §5, for GL(2). □

7.4 Lifting of Jacobi forms

Before discussing the lifting of Jacobi forms to the adele group, we make some general remarks on Fourier expansions. Let Φ be a function on $G^J(F)\backslash G^J(\mathbb{A})$, which for simplicity we assume to be continuous, although the following considerations also make sense, with slight modifications, for more general types of functions, e.g. measurable ones. We assume

$$\Phi(gz) = \psi^m(z)\Phi(g) \qquad \text{for all } g \in G^J(\mathbb{A}),\ z \in Z(\mathbb{A}). \qquad (7.12)$$

For $n, r \in F$ we define the character $\psi^{n,r}$ of $N^J(\mathbb{A})$ as

$$\psi^{n,r}\left(\begin{pmatrix} 1 & x \\ 0 & 1 \end{pmatrix}(0,\mu,0)\right) = \psi(nx + r\mu).$$

If (n, r) runs through F^2, then $\psi^{n,r}$ obviously runs through all characters of $N^J(F)\backslash N^J(\mathbb{A}) \simeq (F\backslash \mathbb{A})^2$.

7.4.1 Definition. For $n, r \in F$, the (n, r)-**Whittaker-Fourier coefficient** of Φ is defined as the continuous function

$$W_\Phi^{n,r}(g) = \int\limits_{N^J(F)\backslash N^J(\mathbb{A})} \Phi(ug)\overline{\psi^{n,r}(u)}\, du,$$

if this integral makes sense.

7.4.2 Lemma. Let $n, r \in F$ and Φ as above.

i) We have

$$W_\Phi^{n,r}(ug) = \psi^{n,r}(u)W_\Phi^{n,r}(g) \qquad \text{for all } g \in G^J(\mathbb{A}),\ u \in N^J(\mathbb{A}).$$

ii) If $a \in F^*$, $\lambda \in F$, then

$$W_\Phi^{n,r}\left(\begin{pmatrix} a & 0 \\ 0 & a^{-1} \end{pmatrix}(\lambda,0,0)g\right) = W_\Phi^{n',r'}(g) \qquad \text{for all } g \in G^J(\mathbb{A}),$$

where $n' = a^2(n + m\lambda^2 + r\lambda)$, $r' = a(r + 2m\lambda)$.

Proof: The first item is obvious, while the second one follows easily from the left $G^J(F)$-invariance of Φ and (7.12). $\qquad\square$

7.4.3 Remark. As the **discriminant** of the pair $(n, r) \in F^2$ we denote the number

$$N = 4mn - r^2.$$

The discriminant of the pair (n', r') with n', r' as in the lemma is then

$$N' = 4mn' - r'^2 = a^2 N.$$

Conversely, given a pair $(n', r') \in F^2$ such that this latter equation holds for some $a \in F^*$, we can define $\lambda = (r'a^{-1} - r)/(2m)$, and the above equations between (n, r) and (n', r') hold. We can thus state the following: If we know $W_\Phi^{n,r}$, then we know $W_\Phi^{n',r'}$ for all pairs $(n', r') \in F^2$ whose discriminant differs from $N = 4mn - r^2$ by a square in F^*. The pairs $(n, r) \in F^2$ (and thus the Whittaker-Fourier coefficients of Φ) are partitioned into orbits, indexed by $N \in F/F^{*2} = \{0\} \cup F^*/F^{*2}$. In order to reconstruct Φ, we have to know at least one Whittaker-Fourier coefficient from each orbit. This is in contrast to the situation for GL(2), where it is well known that an automorphic function is completely determined by its first Fourier coefficient. In our case, the pairs with discriminant 0 make up one orbit.

7.4.4 Definition. *The function Φ as above is called* **cuspidal** *if $W_\Phi^{0,0} = 0$, i.e.*

$$\int\limits_{N^J(F)\backslash N^J(\mathbb{A})} \Phi(ng)\, dn = 0 \qquad \text{for all } g \in G^J(\mathbb{A}).$$

The above remarks show that if Φ is cuspidal, then not only the $(0,0)$-Whittaker-Fourier coefficient vanishes, but all Whittaker-Fourier coefficients with discriminant 0 do also. This is in contrast with the situation in the real case, where there are finitely many conditions to check (cf. the discussion in Section 4.2).

Now we come to what is announced in the title of this section. The decomposition (7.3) allows us to lift classical Jacobi forms, as defined in 4.1.1, to the adele group. The process is familiar from the theory of elliptic modular forms, which are usually lifted to GL(2, $\mathbb{A}$). The number field is from now on $\mathbb{Q}$.

To a Jacobi form $f \in J_{k,m}$ the lift Φ_f is the function $G^J(\mathbb{A}) \to \mathbb{C}$ defined in the following way. According to (7.2), an element $g \in G^J(\mathbb{A})$ may be decomposed as

$$g = \gamma g_\infty k_0 \qquad \text{with } \gamma \in G^J(\mathbb{Q}),\ g_\infty \in G^J(\mathbb{R}),\ k_0 \in K_0^J.$$

Then

$$\Phi_f(g) := \left(f\Big|_{k,m} g_\infty\right)(i, 0) = j_{k,m}(g_\infty, (i,0))f(g_\infty(i,0)).$$

This is well defined since $\Gamma^J = G^J(\mathbb{Q}) \cap K_0^J$.

7.4.5 Proposition. *Under the map $f \mapsto \Phi_f$ the space $J_{k,m}$ is isomorphic to the space $\mathcal{A}_{m,k}$ of functions $\Phi : G^J(\mathbb{A}) \to \mathbb{C}$ having the following properties:*

i) $\Phi(\gamma g) = \Phi(g)$ for all $g \in G^J(\mathbb{A})$, $\gamma \in G^J(\mathbb{Q})$.

ii) $\Phi(gz) = \psi^m(z)\Phi(g)$ for all $g \in G^J(\mathbb{A})$, $z \in Z(\mathbb{A})$.

iii) $\Phi(gk_\infty) = j_{k,m}(k_\infty, (i, 0))\Phi(g)$ for all $g \in G^J(\mathbb{A})$, $k_\infty \in K_\infty$.

iv) $\Phi(gk_0) = \Phi(g)$ for all $g \in G^J(\mathbb{A})$, $k_0 \in K_0^J$.

v) Φ is C^∞ as a function on $G^J(\mathbb{R})$, and as such

$$\mathcal{L}_{X_-}\Phi = \mathcal{L}_{Y_-}\Phi = 0.$$

vi) Φ is slowly increasing.

The condition v) may be substituted by

v') Φ is C^∞ as a function on $G^J(\mathbb{R})$, and as such

$$\mathcal{L}_C\Phi = \left(k^2 - 3k + \frac{5}{4}\right)\Phi.$$

The restriction to $J_{k,m}^{\text{cusp}}$ yields an isomorphism onto the space $\mathcal{A}_{m,k}$ of functions Φ which fulfill i)–vi) and are moreover cuspidal, i.e.

$$\int_{N^J(F)\backslash N^J(\mathbb{A})} \Phi(ng)\, dn = 0 \qquad \text{for all } g \in G^J(\mathbb{A}).$$

Proof: These are routine arguments, except perhaps for the assertion that v) may be substituted by v'). This follows from Corollary 4.3.5, since our lifting may be carried out in two steps: First we lift to $G^J(\mathbb{R})$, obtaining the space described in Corollary 4.3.5, and after that lift these functions to $G^J(\mathbb{A})$ by strong approximation.

We check the cusp condition. Let

$$f(\tau, z) = \sum_{4mn-r^2 \geq 0} c(n, r)e(n\tau + rz)$$

be the Fourier expansion of our classical Jacobi form f. Then by straightforward calculations we have for the Whittaker-Fourier coefficients of the lift Φ

$$W_\Phi^{n,r}\left(\begin{pmatrix} a & 0 \\ 0 & a^{-1} \end{pmatrix}(\lambda, 0, 0)\right) = c(n, r)a^k e^{-2\pi(na^2 + ra\lambda + m\lambda^2)} \quad (a \in \mathbb{R}_{>0}, \lambda \in \mathbb{R}),$$

if $n, r \in \mathbb{Z}$ with $4mn - r^2 \geq 0$, while $W_\Phi^{n,r} = 0$ otherwise. By the remark made after Definition 7.4.4 we see that Φ is cuspidal in the sense of this definition if and only if f is a cusp form. $\qquad\square$

7.4.6 Proposition. $\mathcal{A}_{m,k}^{\mathrm{cusp}}$ *is a subspace of* $L_0^2(G^J(\mathbb{Q})\backslash G^J(\mathbb{A}))_m$, *and* $f \mapsto \Phi_f$ *is an isomorphism of Hilbert spaces* $J_{k,m}^{\mathrm{cusp}} \to \mathcal{A}_{m,k}^{\mathrm{cusp}}$.

Proof: This follows from (7.6) in Section 7.1 and the fact that the classical Petersson scalar product in equation (13) on page 27 of [EZ] is finite for cusp forms. $\qquad\square$

Let $\Phi : G^J(\mathbb{A}) \to \mathbb{C}$ be the lift of a Jacobi form f with weight k and index m. If f is not a cusp form, then it need not be true that $\Phi \in L^2(G^J(\mathbb{Q})\backslash G^J(\mathbb{A}))_m$. Nevertheless, a slight modification of Lemma 7.3.2 (leaving out the L^2-conditions) shows that Φ can be written in the form

$$\Phi(Mh) = \sum_i \varphi_i(M, 1)\vartheta_{f_i}(Mh) \tag{7.13}$$

with genuine functions $\varphi_i : \mathrm{Mp}(\mathbb{Q})\backslash\mathrm{Mp}(\mathbb{A}) \to \mathbb{C}$ and Schwartz functions f_i in $\mathcal{S}(\mathbb{A})$. We can derive information on the f_i from this equation, letting the Heisenberg group act on both sides, which does not affect the φ_i. Let us begin with the infinite place. The holomorphy of f implies $Y_-\Phi = 0$. Assuming the functions φ_i linearly independent, it follows that $Y_-\vartheta_{f_i} = 0$, or equivalently, $Y_-f_i = 0$, for all i. But the real Schrödinger representation contains (up to scalars) only one vector annihilated by Y_-, namely the function

$$F_\infty(x) = e^{-2\pi m x^2} \in \mathcal{S}(\mathbb{R})$$

(remember in the context of Jacobi forms we assume $m > 0$). Hence each f_i may be assumed to be of the form

$$f_i = F_\infty \otimes F_i \qquad \text{with } F_i \in \mathcal{S}(\mathbb{A}_0).$$

Now we continue to get information about the F_i by letting the finite parts of the Heisenberg group act on (7.13). The left side is invariant under $H(\hat{\mathbb{Z}})$, and again the linear independence of the φ_i implies that each ϑ_{f_i}, and therefore each F_i, is right invariant under $H(\hat{\mathbb{Z}})$. Now from Lemma 6.3.2 one can read off a basis for the $H(\mathbb{Z}_p)$-invariant vectors in $\mathcal{S}(\mathbb{Q}_p)$ of the local Schrödinger representation $\pi_{SW,p}^m$; it is given by

$$f_{p,\nu} = \mathrm{char}\Big(\mathbb{Z}_p + \frac{\nu}{2m}\Big), \qquad \nu \in \mathbb{Z}_p/2m\mathbb{Z}_p.$$

The $H(\hat{\mathbb{Z}})$-invariant vectors in $\mathcal{S}(\mathbb{A}_0)$ therefore have a basis consisting of the $2m$ elements

$$f_\nu = \bigotimes_{p<\infty} \mathrm{char}\Big(\mathbb{Z}_p + \frac{\nu}{2m}\Big), \qquad \nu \in \mathbb{Z}/2m\mathbb{Z}.$$

Each F_i is a linear combination of these functions. Summing up, we may assume that

$$\Phi(Mh) = \sum_{\nu\in\mathbb{Z}/2m\mathbb{Z}} \varphi_\nu(M, 1)\vartheta_\nu(Mh), \tag{7.14}$$

where the φ_ν are certain genuine functions on $\mathrm{Mp}(\mathbb{Q})\backslash\mathrm{Mp}(\mathbb{A})$, and ϑ_ν is the theta function corresponding to $F_\infty \otimes f_\nu \in \mathcal{S}(\mathbb{A})$. The relation (7.14) is nothing else than a lifted version of equation (5) on page 58 of [EZ]. Just like there we can deduce some properties of the functions φ_ν from known properties of the ϑ_ν. For example, from

$$\vartheta_\nu(M(-\lambda,-\mu,\kappa)) = \vartheta_{-\nu}(M(\lambda,\mu,\kappa)) \qquad \text{for all } M \in G(\mathbb{A}),\ \lambda,\mu,\kappa \in \mathbb{A},$$

which is easily seen, and

$$\Phi(M(\lambda,\mu,\kappa)) = \Phi((-1)M(\lambda,\mu,\kappa))$$
$$= \Phi(M(-\lambda,-\mu,\kappa)(-1)) = (-1)^k\Phi(M(\lambda,\mu,\kappa)),$$

we deduce

$$\varphi_{-\nu} = (-1)^k\varphi_\nu.$$

With

$$\tilde{\vartheta}_\nu := \begin{cases} \vartheta_\nu + (-1)^k\vartheta_{-\nu} & \text{if } \nu \not\equiv 0, m \mod 2m, \\ \vartheta_\nu & \text{if } \nu \equiv 0, m \mod 2m, \end{cases}$$

one can therefore write

$$\Phi(Mh) = \sum_{\nu=0}^{m} \varphi_\nu(M,1)\tilde{\vartheta}_\nu(Mh) \tag{7.15}$$

(the terms $\nu = 0$ and $\nu = m$ vanish for odd k). There is a stronger symmetry property (see Proposition 7.4.9) if we require f to be an eigenfunction for certain Hecke operators at the bad places, which we introduce now.

The Heisenberg involutions

The (classical analog of) decomposition (7.14) is used in [EZ], Theorem 5.2, to construct an involutive automorphism W_p of $J_{k,m}$, for each prime p dividing m, which commutes with all Hecke operators $T_{EZ}(n)$, $(n,m) = 1$. We will give now an 'explanation' of these W_p in terms of certain elements of the local Hecke algebra $\mathcal{H}(G^J(\mathbb{Q}_p), G^J(\mathbb{Z}_p))$. Namely, what we will show is that W_p coincides with the action of the element

$$B_p := p^{-2l} \sum_{\lambda,\mu,\kappa\in\mathbb{Z}/p^l\mathbb{Z}} \mathrm{char}\Big(G^J(\mathbb{Z}_p)(\lambda p^{-l}, \mu p^{-l}, \kappa p^{-l})\Big) \tag{7.16}$$

of $\mathcal{H}(G^J(\mathbb{Q}_p), G^J(\mathbb{Z}_p))$, where $l = v_p(m)$. Note that the disjoint union of the cosets $G^J(\mathbb{Z}_p)(\lambda p^{-l}, \mu p^{-l}, \kappa p^{-l})$ is right $G^J(\mathbb{Z}_p)$-invariant, and thus is indeed a union of double cosets $G^J(\mathbb{Z}_p)gG^J(\mathbb{Z}_p)$ with $g \in G^J(\mathbb{Q}_p)$ (it consists of exactly

$l + 1$ such double cosets). It is not true that $B_p^2 = 1$ in $\mathcal{H}(G^J(\mathbb{Q}_p), G^J(\mathbb{Z}_p))$. But by some routine Hecke algebra calculations one can show that

$$\Xi(B_p)^2 = 1 \qquad \text{in } \mathcal{H}(G^J(\mathbb{Q}_p), G^J(\mathbb{Z}_p), \psi_p^{-m}),$$

where Ξ denotes the natural homomorphism

$$\mathcal{H}(G^J(\mathbb{Q}_p), G^J(\mathbb{Z}_p)) \longrightarrow \mathcal{H}(G^J(\mathbb{Q}_p), G^J(\mathbb{Z}_p), \psi_p^{-m})$$

from (6.1). Since the action of $\mathcal{H}(G^J(\mathbb{Q}_p), G^J(\mathbb{Z}_p))$ factors through the Hecke algebra $\mathcal{H}(G^J(\mathbb{Q}_p), G^J(\mathbb{Z}_p), \psi_p^{-m})$, we see that B_p operates as an involution on the space $\mathcal{A}_{m,k}$ of lifted Jacobi forms, and so does it on $J_{k,m}$. We call this operator the p-th **Heisenberg involution**, and will compute its action now. So let Φ be the adelic lift of the Jacobi form f, and write Φ in the form (7.14). Let

$$2m = \prod_{i=1}^{r} p_i^{\alpha_i}$$

be the prime factor decomposition of the natural number $2m$, where we set $p_1 = 2$. We choose integers a_i $(i = 1, \ldots, r)$ such that

$$a_i \equiv 1 \quad \text{mod } p_i^{\alpha_i},$$
$$a_j \equiv 0 \quad \text{mod } p_j^{\alpha_j} \qquad \text{for all } j \neq i.$$

If ν_i runs through a system of representatives of $\mathbb{Z}/p_i^{\alpha_i}\mathbb{Z}$, then, by the Chinese remainder theorem,

$$\nu = \sum_{i=1}^{r} a_i \nu_i$$

runs through a system of representatives of $\mathbb{Z}/2m\mathbb{Z}$. The decomposition (7.14) shall accordingly be written in the form

$$\Phi(Mh) = \sum_{\nu_1} \cdots \sum_{\nu_r} \varphi_{\nu_1,\ldots,\nu_r}(M,1)\vartheta_{\nu_1,\ldots,\nu_r}(Mh). \tag{7.17}$$

Here ν_i runs through an arbitrary complete system of representatives of $\mathbb{Z}/p_i^{\alpha_i}\mathbb{Z}$, but we may think of ν_i as to run from 0 to $p_i^{\alpha_i} - 1$, the latter number being odd for $i = 1$ and even otherwise. Since the Hecke operator B_p has pure Heisenberg elements as coset representatives, the action of B_{p_i} on Φ has no effect on the functions $\varphi_{\nu_1,\ldots,\nu_r}$, but only on the theta functions $\vartheta_{\nu_1,\ldots,\nu_r}$ (we let $l := v_{p_i}(m)$, which differs from α_i only in the case $2|m$ and $i = 1$):

$$
(\Phi * B_{p_i})(Mh) = p_i^{-2l} \sum_{\lambda,\mu,\kappa \in \mathbb{Z}/p_i^l \mathbb{Z}} \Phi(Mh(\lambda p_i^{-l}, \mu p_i^{-l}, \kappa p_i^{-l}))
$$

$$
= p_i^{-2l} \sum_{\nu_1} \cdots \sum_{\nu_r} \sum_{\lambda,\mu,\kappa \in \mathbb{Z}/p_i^l \mathbb{Z}} \varphi_{\nu_1,\ldots,\nu_r}(M,1)
$$
$$
\vartheta_{\nu_1,\ldots,\nu_r}(Mh(\lambda p_i^{-l}, \mu p_i^{-l}, \kappa p_i^{-l}))
$$

$$
= \sum_{\nu_1} \cdots \sum_{\nu_r} \varphi_{\nu_1,\ldots,\nu_r}(M,1)(\vartheta_{\nu_1,\ldots,\nu_r} * B_{p_i})(Mh).
$$

Note that in this computation, the elements $(\lambda p_i^{-l}, \mu p_i^{-l}, \kappa p_i^{-l})$ are in $H(\mathbb{Q}_{p_i})$, while M and h are global elements.

7.4.7 Lemma. *With the above notations, we have*

$$
\vartheta_{\nu_1,\ldots,\nu_r} * B_{p_i} = \vartheta_{\nu_1,\ldots,-\nu_i,\ldots,\nu_r} \qquad \text{for all } i \in \{1,\ldots,r\}.
$$
$$
(7.18)
$$

Proof: By definition, $\vartheta_{\nu_1,\ldots,\nu_r}$ corresponds to a global Schwartz function whose p_i-component is

$$
f_{p_i,\nu_1,\ldots,\nu_r} = \mathrm{char}\left(\mathbb{Z}_{p_i} + \frac{\nu_i}{2m}\right).
$$

From now on we simplify the notation by skipping the index i. The central character ψ_p^m of the local Schrödinger representation at the place p is trivial on $p^{-l}\mathbb{Z}_p$, hence

$$
(\vartheta_{\nu_1,\ldots,\nu_r} * B_p)(Mh) = p^{-2l} \sum_{\lambda,\mu,\kappa \in \mathbb{Z}/p^l \mathbb{Z}} \vartheta_{\nu_1,\ldots,\nu_r}(Mh(\lambda p^{-l}, \mu p^{-l}, \kappa p^{-l}))
$$

$$
= p^{-l} \sum_{\lambda,\mu \in \mathbb{Z}/p^l \mathbb{Z}} \vartheta_{\nu_1,\ldots,\nu_r}(Mh(\lambda p^{-l}, \mu p^{-l}, 0))
$$

From

$$
\vartheta_{\nu_1,\ldots,\nu_r}(Mh) = \sum_{\xi \in \mathbb{Q}} \left(\prod_{p' \leq \infty} \left(\pi_{SW,p'}^m(M_{p'} h_{p'}) f_{p',\nu_1,\ldots,\nu_r} \right) \right)(\xi)
$$

we see that we have to compute

$$
\sum_{\lambda,\mu \in \mathbb{Z}/p^l \mathbb{Z}} \left(\pi_{S,p}^m(\lambda p^{-l}, \mu p^{-l}, 0) f_{p,\nu_1,\ldots,\nu_r} \right)(x)
$$

$$
= \sum_{\lambda,\mu \in \mathbb{Z}/p^l \mathbb{Z}} \psi_p^m((2x + \lambda p^{-l})\mu p^{-l}) f_{p,\nu_1,\ldots,\nu_r}(x + \lambda p^{-l}).
$$

From this it is clear that if $x \notin \frac{1}{2m}\mathbb{Z}_p$, then both sides of (7.18) are zero. Assume conversely that $x \in \frac{1}{2m}\mathbb{Z}_p$. Then the sum over μ gives 0 unless $2x + \lambda p^{-l} \in \mathbb{Z}_p$, in which case the argument of ψ_p lies in $\mathbb{Z}_p$. Hence

$$\sum_{\lambda,\mu \in \mathbb{Z}/p^l\mathbb{Z}} \left(\pi^m_{S,p}(\lambda p^{-l}, \mu p^{-l}, 0) f_{p,\nu_1,\ldots,\nu_r} \right)(x)$$

$$= \sum_{\substack{\lambda \in \mathbb{Z}/p^l\mathbb{Z} \\ 2x+\lambda p^{-l} \in \mathbb{Z}_p}} p^l f_{p,\nu_1,\ldots,\nu_r}(x + \lambda p^{-n})$$

$$= p^l f_{p,\nu_1,\ldots,\nu_r}(-x) = p^l f_{p,\nu_1,\ldots,-\nu_i,\ldots,\nu_r}(x).$$

This proves our assertion. Note that this latter computation remains valid for $i = 1$, i.e. $p = 2$, if $2|m$. $\qquad\square$

We see that the action of our operator B_p on Φ amounts to just altering the way the theta functions ϑ_ν are attached to the half-integer modular forms φ_ν, and then summing up over ν to get a Jacobi form. To be precise, if we choose an integer $u \in \mathbb{Z}$ such that

$$u \equiv 1 \quad \mathrm{mod}\ 2m/m', \qquad u \equiv -1 \quad \mathrm{mod}\ 2m',$$

where $m' = p^{v_p(m)}$ is the p-part of m, then we have shown that

$$(\Phi * B_p)(Mh) = \sum_{\nu \in \mathbb{Z}/2m\mathbb{Z}} \varphi_\nu(M)\vartheta_{u\nu}(Mh).$$

But this is precisely the way in which the operators W_p on page 60 of [EZ] are constructed. To summarize:

7.4.8 Proposition. *For $p|m$ let B_p be the Heisenberg involution defined by (7.16), and let W_p be the involution on $J_{k,m}$ defined on page 60 of [EZ]. Then for any $f \in J_{k,m}$ with adelic lift Φ_f*

$$\Phi_f * B_p = \Phi_{W_p f}.$$

Since operators which come from different local Hecke algebras are commuting, we see that $J_{k,m}$ has a basis consisting of eigenforms for all $T^J(p)$, $p \nmid m$, and all B_p, $p|m$. Such an eigenform has the following property.

7.4.9 Proposition. *Assume $f \in J_{k,m}$ is an eigenform for all Heisenberg involutions B_p, $p|m$. Then the adelic lift Φ_f has a decomposition as a finite sum*

$$\Phi_f(Mh) = \sum_i \varphi_i(M,1)\vartheta_i(Mh)$$

where φ_i are functions on the metaplectic group as before, and where every ϑ_i is a theta function corresponding to a Schwartz function $F_i \in \mathcal{S}(\mathbb{A})$ with the following property: F_i is a pure tensor

$$F_i = \bigotimes_{p' \leq \infty} F_{i,p'},$$

and for every finite p', the local component $F_{i,p'}$ lies either in the space $\mathcal{S}(F)^+$ of even or in the space $\mathcal{S}(F)^-$ of odd Schwartz functions, and this independently of i.

Proof: We start with the decomposition (7.17),

$$\Phi_f(Mh) = \sum_{\nu_1} \cdots \sum_{\nu_r} \varphi_{\nu_1,\dots,\nu_r}(M,1)\vartheta_{\nu_1,\dots,\nu_r}(Mh). \tag{7.19}$$

We have to distinguish between the cases $2|m$ and $2 \nmid m$. Since they are treated very similarly, and differ mainly in the notation, we only treat the latter one, and leave the minor changes for the former one to the reader. Hence assume $2 \nmid m$. Then, with the notations used before, $\alpha_1 = 1$, and $p_2,\dots,p_r$ are the prime divisors of m. Now, for any $i \in \{2,\dots,r\}$, since f is an eigenfunction under the involution W_{p_i}, by the preceding proposition there exists a sign $\varepsilon_i \in \{\pm 1\}$ such that

$$\Phi_f * B_{p_i} = \varepsilon_i \Phi_f.$$

On the other hand, we have computed above that

$$(\Phi_f * B_{p_i})(Mh) = \sum_{\nu_1} \cdots \sum_{\nu_r} \varphi_{\nu_1,\dots,\nu_r}(M,1)\vartheta_{\nu_1,\dots,-\nu_i,\dots,\nu_r}(Mh). \tag{7.20}$$

Since the theta functions are linearly independent, we deduce the symmetry property

$$\varphi_{\nu_1,\dots,-\nu_i,\dots,\nu_r} = \varepsilon_i\varphi_{\nu_1,\dots,\nu_r} \qquad \text{for all } i \in \{2,\dots,r\} \text{ and all } \nu_j. \tag{7.21}$$

Since $2 \nmid m$, the number ν_1 in the above sum just takes on the values 0 and 1. The local component of the theta function $\vartheta_{\nu_1,\dots,\nu_r}$ at the place 2 is accordingly

$$f_{2,\nu_1,\dots,\nu_r} = \text{char}(\mathbb{Z}_2)$$

or

$$f_{2,\nu_1,\dots,\nu_r} = \text{char}\left(\mathbb{Z}_2 + \frac{1}{2m}\right).$$

In any case, $f_{2,\nu_1,\dots,\nu_r}$ is an even Schwartz function. Now we look at the next local component $f_{p_2,\nu_1,\dots,\nu_r}$. The index ν_2 may be assumed to run through the set

$$\left\{-\frac{p_2^{\alpha_2}-1}{2},\dots,\frac{p_2^{\alpha_2}-1}{2}\right\}.$$

If we define

$$
\vartheta^{(2)}_{\nu_1,\nu_2,\dots,\nu_r} = \begin{cases} \vartheta_{\nu_1,\nu_2,\dots,\nu_r} + \varepsilon_2\vartheta_{\nu_1,-\nu_2,\dots,\nu_r} & \text{for } \nu_2 \in \left\{1,\dots,\frac{p_2^{\alpha_2}-1}{2}\right\}, \\ \vartheta_{\nu_1,0,\dots,\nu_r} & \text{for } \nu_2 = 0, \end{cases}
$$

then, because of the symmetry property (7.21), we can write

$$
\Phi_f(Mh) = \sum_{\nu_1}\sum_{\nu_2=0}^{(p_2^{\alpha_2}-1)/2}\cdots\sum_{\nu_r}\varphi_{\nu_1,\dots,\nu_r}(M,1)\vartheta^{(2)}_{\nu_1,\dots,\nu_r}(Mh). \tag{7.22}
$$

Note that if $\varepsilon_2 = -1$, then $\varphi_{\nu_1,0,\dots,\nu_r} = 0$, and the second sum could start with $\nu_2 = 1$. Hence we see that each $\vartheta^{(2)}_{\nu_1,\dots,\nu_r}$ which really occurs corresponds to a Schwartz function which is a pure tensor, and whose local component at the place p_2 lies in the space of even (if $\varepsilon_2 = 1$), resp. odd (if $\varepsilon_2 = -1$), Schwartz functions. Now we go on like this and define inductively

$$
\vartheta^{(3)}_{\nu_1,\nu_2,\dots,\nu_r} = \begin{cases} \vartheta^{(2)}_{\nu_1,\nu_2,\nu_3,\dots,\nu_r} + \varepsilon_3\vartheta^{(2)}_{\nu_1,\nu_2,-\nu_3,\dots,\nu_r} & \text{for } \nu_3 \in \left\{1,\dots,\frac{p_3^{\alpha_3}-1}{2}\right\}, \\ \vartheta^{(2)}_{\nu_1,0,\dots,\nu_r} & \text{for } \nu_3 = 0, \end{cases}
$$

$$
\vdots
$$

$$
\vartheta^{(r)}_{\nu_1,\nu_2,\dots,\nu_r} = \begin{cases} \vartheta^{(r-1)}_{\nu_1,\nu_2,\dots,\nu_r} + \varepsilon_3\vartheta^{(r-1)}_{\nu_1,\nu_2,\dots,-\nu_r} & \text{for } \nu_r \in \left\{1,\dots,\frac{p_r^{\alpha_r}-1}{2}\right\}, \\ \vartheta^{(r-1)}_{\nu_1,0,\dots,\nu_r} & \text{for } \nu_r = 0, \end{cases}
$$

Then we have

$$
\Phi_f(Mh) = \sum_{\nu_1}\sum_{\nu_2=0}^{(p_2^{\alpha_2}-1)/2}\cdots\sum_{\nu_r=0}^{(p_r^{\alpha_r}-1)/2}\varphi_{\nu_1,\dots,\nu_r}(M,1)\vartheta^{(r)}_{\nu_1,\dots,\nu_r}(Mh), \tag{7.23}
$$

and this is a decomposition as desired. $\square$

7.4.10 Remark. The purpose of the decomposition established in this proposition is as follows. Assume a global matrix

$$
M = (M_p)_{p\leq\infty}
$$

where each M_p is plus or minus the identity matrix. Then there is a $\varepsilon \in \{\pm1\}$ such that

$$
\vartheta_i(gM) = \varepsilon\vartheta_i(g) \qquad\text{for all } i \text{ and all } g \in G(\mathbb{A}).
$$

This is because right translation with M corresponds to applying $\pi_w^m(M)$ to the corresponding Schwartz function, and the local components of these Schwartz functions lie in irreducible subspaces for the local Weil representations (namely, the space of even, resp. odd, Schwartz functions). The following section will explain the significance of this property for the problem of proving the irreducibility of the representation which can be assigned to a Jacobi form.

7.5 The representation corresponding to a Jacobi form

In this section we want to establish a connection between classical Jacobi cusp forms in $J_{k,m}^{\text{cusp}}$ and automorphic representations of G^J. To explain the procedure, we first recall the analogous situation for ordinary elliptic modular forms, following [Ge1], §3 and §5.

Let $f \in S_k(\Gamma_0(m))$ be a classical cusp form of weight k and level m. We want to associate to f an automorphic representation π_f of $\mathrm{GL}(2, \mathbb{A})$, where $\mathbb{A}$ is the adele ring of $\mathbb{Q}$. The first step is to lift f to a function Φ_f on $\mathrm{GL}(2, \mathbb{A})$, just as we lifted Jacobi forms to functions on $G^J(\mathbb{A})$ (Proposition 7.4.5). Since f is a cusp form, this Φ_f will lie in the Hilbert space

$$\mathcal{H} = L^2(\mathrm{GL}(2, \mathbb{Q}) \backslash \mathrm{GL}(2, \mathbb{A}), \mathbf{1})$$

of measurable, square integrable functions on $\mathrm{GL}(2, \mathbb{Q}) \backslash \mathrm{GL}(2, \mathbb{A})/Z$, where Z denotes the center of $\mathrm{GL}(2, \mathbb{A})$ ("$\mathbf{1}$" stands for the trivial character of Z). One defines π_f as the subrepresentation of $\mathcal{H}$ generated by Φ_f. The question then is: Is this π_f irreducible?

It turns out that π_f is irreducible if one requires f to be a Hecke eigenform, i.e., an eigenvector under all Hecke operators $T(p)$ at the good primes $p \nmid m$. The proof is as follows. The unitary representation π_f is decomposed into irreducibles:

$$\pi_f = \bigoplus_i \pi_i$$

with irreducible subrepresentations π_i of $\mathcal{H}$. Each of the π_i is decomposed into a restricted tensor product of local representations:

$$\pi_i = \bigotimes_{p \leq \infty} \pi_{i,p}.$$

It is rather easy to see that the local components $\pi_{i,\infty}$ at the infinite place agree for all i. The same is true for the finite places which do not divide the level m, as a consequence of f being a Hecke eigenform. Now one invokes a *strong multiplicity-one theorem* for $\mathrm{GL}(2)$: Two automorphic representations of $\mathrm{GL}(2, \mathbb{A})$ coincide if (and only if) their local components are isomorphic for the infinite place and almost all finite places. It follows that the π_i all agree, or in other words, that there is only one i. Thus π_f is irreducible.

We summarize the steps which lead from a classic modular form to an automorphic representation:

- Lift the eigenform f to a function Φ_f on $GL(2, \mathbb{A})$.

- Define π_f as the smallest invariant subspace of the Hilbert space $\mathcal{H}$ which contains Φ_f.

- Decompose π_f into irreducibles, and show, by the help of a strong multiplicity-one theorem, that the irreducible components all coincide.

We make two more comments on the group underlying elliptic modular forms. In the above procedure we have modelled them as functions on $GL(2, \mathbb{A})$ which transform trivially under the center. Accordingly, what we are really dealing with is representations of the group $PGL(2, \mathbb{A})$. But if we had considered a modular form with character $f \in S_k(\Gamma_0(m), \chi)$, then the action of the center would have been non-trivial, and the passage to the projective group not possible.

The second remark is about $SL(2)$, which might also appear as a natural domain for elliptic modular forms. But for $SL(2)$, there is no such multiplicity-one result as there is for $GL(2)$. Hence the above procedure to map modular forms into automorphic $SL(2)$-representations would fail. This is reason enough to consider $GL(2)$ instead of $SL(2)$; others are given on page 49 of [Gel]. The usual procedure (reproduced in [Gel] §2) to realize modular forms on the Lie group $SL(2, \mathbb{R})$ and to make the connection with the representation theory of this group is somewhat misleading in the automorphic context.

Now we turn back to Jacobi forms. A cuspidal Jacobi form $f \in J_{k,m}^{\mathrm{cusp}}$ can be regarded as an element Φ_f of the Hilbert space $L_0^2(G^J(\mathbb{Q})\backslash G^J(\mathbb{A}))_m$, cf. Proposition 7.4.6. We assume f to be an eigenform, and we would like to prove that the representation generated by Φ_f is irreducible. This will turn out to be true, provided we have the correct notion of "eigenform".

The above considerations show that what we need as a basic ingredient is a strong multiplicity-one result for the Jacobi group. By corollary 7.3.5, this is equivalent to a strong multiplicity-one result for the metaplectic group. Indeed, such a theorem was discovered by Waldspurger. It is not as smooth as the corresponding $GL(2)$-theorem, but involves conditions at *all* places.

7.5.1 Theorem. *Two cuspidal automorphic representations of* Mp *are identical if and only if they have the same central character and almost all of their local components are isomorphic.*

Proof: See Théorème 3 in [Wa3] or 1.4 in [GePS3]. $\square$

Let S be the finite set of places where two automorphic Mp-representations π_1 and π_2 may fail to be isomorphic. The theorem asserts that $\pi_1 = \pi_2$ if and only if at the finitely many places $\mathfrak{p} \in S$ the central characters of the local components coincide. Now the center of every local group $\mathrm{Mp}(F_{\mathfrak{p}})$ consists of the four elements

$$(\pm 1, \pm 1).$$

Since we are only considering genuine representations, there are exactly two possibilities for the central character of an irreducible representation of $\mathrm{Mp}(F_{\mathfrak{p}})$: The element $(-1, 1)$ can act as $+1$ or -1. We will prove in a moment that for the corresponding G^J-representation this value is connected with the eigenvalue of the Heisenberg involution on a spherical vector. This will mean that in our lifting procedure, we can apply the multiplicity-one theorem 7.5.1 provided the Jacobi form we started with is an eigenfunction for the Hecke operators $T^J(p)$, $p \nmid m$, and the Heisenberg involutions W_p, $p|m$. This is the correct notion of eigenform for Jacobi forms.

To prove the result mentioned, we will need a local version of Lemma 7.4.7. We fix a $\mathfrak{p}$-adic field F, an additive character $\psi : F \to \mathbb{C}$ with conductor $\mathcal{O}$ (the ring of integers in F), and an element $m \in F^*$ with valuation $n = v(m) \geq 0$. In somewhat greater generality than (7.16), we define the element

$$B := q^{-2n} \sum_{\lambda, \mu, \kappa \in \mathcal{O}/\omega^n \mathcal{O}} \mathrm{char}\Big(G^J(\mathcal{O})(\lambda \omega^{-n}, \mu \omega^{-n}, \kappa \omega^{-n})\Big) \qquad (7.24)$$

of $\mathcal{H}(G^J(F), G^J(\mathcal{O}))$, where ω is a prime element and q the cardinality of $\mathcal{O}/\omega\mathcal{O}$. The natural homomorphism

$$\Xi : \mathcal{H}(G^J(F), G^J(\mathcal{O})) \longrightarrow \mathcal{H}(G^J(F), G^J(\mathcal{O}), \psi^{-m})$$

from (6.1) sends B to an element of order 2, and hence B acts with eigenvalues ± 1 on the space of K^J-invariant vectors in any representation of $G^J(F)$ with central character ψ^m. But first we consider the action of B on a certain vector in the Schrödinger-Weil representation.

7.5.2 Lemma. *Consider the local Schrödinger-Weil representation π_{SW}^m on the standard space $\mathcal{S}(F)$. With $n = v(m)$ as above and $n' = v(2m)$ we have*

$$\pi_{SW}^m(B)\mathbf{1}_{u\omega^{-n'} + \mathcal{O}} = \mathbf{1}_{-u\omega^{-n'} + \mathcal{O}} \qquad \text{for all } u \in \mathcal{O}.$$

Proof: We could refer to the proof of Lemma 7.4.7, but nevertheless carry out the similar computations for clarity. One computes

$$(\pi_{SW}^m(B)\mathbf{1}_{u\omega^{-n'} + \mathcal{O}})(x)$$

$$= q^{-2n} \sum_{\lambda, \mu, \kappa \in \mathcal{O}/\omega^n \mathcal{O}} \Big(\pi_S^m(\lambda \omega^{-n}, \mu \omega^{-n}, \kappa \omega^{-n})\mathbf{1}_{u\omega^{-n'} + \mathcal{O}}\Big)(x)$$

$$= q^{-n} \sum_{\lambda, \mu \in \mathcal{O}/\omega^n \mathcal{O}} \Big(\pi_S^m(\lambda \omega^{-n}, \mu \omega^{-n}, 0)\mathbf{1}_{u\omega^{-n'} + \mathcal{O}}\Big)(x)$$

$$= q^{-n} \sum_{\lambda, \mu \in \mathcal{O}/\omega^n \mathcal{O}} \psi(\mu m \omega^{-n}(2x + \lambda \omega^{-n}))\mathbf{1}_{u\omega^{-n'} + \mathcal{O}}(x + \lambda \omega^{-n}).$$

If $x \notin \omega^{-n'}\mathcal{O}$, then $(\pi_{SW}^m(B)\mathbf{1}_{u\omega^{-n'} + \mathcal{O}})(x) = \mathbf{1}_{-u\omega^{-n'} + \mathcal{O}}(x)$ since both sides equal zero. Assume $x \in \omega^{-n'}\mathcal{O}$. Then $\sum_{\mu \in \mathcal{O}/\omega^n \mathcal{O}} \psi(\mu m \omega^{-n}(2x + \lambda \omega^{-n}))$ is different from zero if and only if $2x + \lambda \omega^{-n} \in \mathcal{O}$, i.e., if and only if λ takes the value $-2x\omega^n$ (note this is an element of $\mathcal{O}$). Hence

$$
\begin{aligned}
(\pi_{SW}^m(B)\mathbf{1}_{u\omega^{-n'}+\mathcal{O}})(x) &= \mathbf{1}_{u\omega^{-n'}+\mathcal{O}}(x - 2x\omega^n\omega^{-n}) \\
&= \mathbf{1}_{u\omega^{-n'}+\mathcal{O}}(-x) = \mathbf{1}_{-u\omega^{-n'}+\mathcal{O}}(x). \qquad \square
\end{aligned}
$$

Let V be the space of an irreducible representation of $G^J(F)$. As an operator on the subspace V^{K^J} of K^J-invariant vectors, B fulfills $B^2 = 1$. Consequently V^{K^J} splits into the ± 1-Eigenspaces $V_{\pm}^{K^J}$:

$$
V^{K^J} = V_+^{K^J} \oplus V_-^{K^J}.
$$

Indeed, only one of the eigenvalues occurs:

7.5.3 Proposition. *Let $\tilde{\pi}$ be an irreducible, admissible (genuine) representation of $\mathrm{Mp}(F)$, and $\pi = \pi_{SW}^m \otimes \tilde{\pi}$ the corresponding representation of $G^J(F)$. Let λ be the central character of $\tilde{\pi}$. Then any spherical vector of π has the B-eigenvalue $\delta_m(-1)\lambda(-1)$. In other words,*

$$
V^{K^J} = V_{\varepsilon}^{K^J} \qquad\qquad \text{with } \varepsilon = \delta_m(-1)\lambda(-1),
$$

where V denotes the space of π.

Proof: A spherical vector v of π can be written in the form

$$
v = \sum_i f_i \otimes \varphi_i
$$

with f_i in the standard space $\mathcal{S}(F)$ of π_{SW}^m, and φ_i in any model of $\tilde{\pi}$. We may assume the f_i as well as the φ_i are linearly independent vectors. The spherical vector v is in particular invariant under $H(\mathcal{O})$, and since the Heisenberg group only acts on the f_i and not on the φ_i, it follows that every f_i is invariant under $H(\mathcal{O})$ (which acts by the Schrödinger representation). Lemma 6.3.2 then tells us the general shape of the f_i. We may thus assume that

$$
v = \sum_{u \in \mathcal{O}/\omega^{n'}\mathcal{O}} \mathbf{1}_{u\omega^{-n'}+\mathcal{O}} \otimes \varphi_u \qquad \text{with } n' = v(2m)
$$

and certain vectors φ_u in the space of $\tilde{\pi}$. From Lemma 7.5.2 we have

$$
\begin{aligned}
\pi(B)v &= \sum_{u \in \mathcal{O}/\omega^{n'}} \left(\pi_{SW}^m(B)\mathbf{1}_{u\omega^{-n'}+\mathcal{O}} \right) \otimes \varphi_u \\
&= \sum_{u \in \mathcal{O}/\omega^{n'}} \mathbf{1}_{-u\omega^{-n'}+\mathcal{O}} \otimes \varphi_u \\
&= \sum_{u \in \mathcal{O}/\omega^{n'}} \mathbf{1}_{u\omega^{-n'}+\mathcal{O}} \otimes \varphi_{-u}.
\end{aligned}
$$

Consequently, for $\varepsilon \in \{-1, 1\}$,

$$\pi(B)v = \varepsilon v \quad \text{is equivalent with} \quad \varphi_u = \varepsilon \varphi_{-u} \quad \text{for all } u \in \mathcal{O}/\omega^{n'}. \tag{7.25}$$

Since v is spherical, we also have

$$
\begin{aligned}
v &= \pi(-1)v = \sum_{u \in \mathcal{O}/\omega^{n'}} \pi_W^m(-1) 1_{u\omega^{n'} + \mathcal{O}} \otimes \tilde{\pi}(-1)\varphi_u \\
&= \delta_m(-1)\lambda(-1) \sum_{u \in \mathcal{O}/\omega^{n'}} 1_{-u\omega^{n'} + \mathcal{O}} \otimes \varphi_u,
\end{aligned}
$$

and as a consequence

$$\varphi_u = \delta_m(-1)\lambda(-1)\varphi_{-u} \qquad \text{for all } u \in \mathcal{O}/\omega^{n'}. \tag{7.26}$$

From (7.25) and (7.26) the assertion follows. □

Theorem 7.5.1 together with Proposition 7.5.3 immediately yields the following strong multiplicity-one result for automorphic representations of the Jacobi group.

7.5.4 Theorem. *Let $\pi_1 = \otimes \pi_{1,\mathfrak{p}}$ and $\pi_2 = \otimes \pi_{2,\mathfrak{p}}$ be automorphic representations of G^J over any number field. Assume there exists a finite set S of finite places such that*

i) $\pi_{1,\mathfrak{p}} \simeq \pi_{2,\mathfrak{p}}$ for all $\mathfrak{p} \notin S$.

ii) For every $\mathfrak{p} \in S$, the representations $\pi_{1,\mathfrak{p}}$ and $\pi_{2,\mathfrak{p}}$ are spherical, and the Heisenberg involution $B_{\mathfrak{p}}$ has the same value on the spaces of K^J-invariant vectors of $\pi_{1,\mathfrak{p}}$ and $\pi_{2,\mathfrak{p}}$.

Then $\pi_1 = \pi_2$.

This is the kind of result which is needed to associate classical Jacobi forms with automorphic representations of $G^J(\mathbb{A})$, the number field being $\mathbb{Q}$. It is almost obvious now how to prove the following theorem.

7.5.5 Theorem. *Let $f \in J_{k,m}^{\text{cusp}}$ be a Hecke eigenform, i.e, an eigenvector for all W_p, $p|m$, and all $T^J(p)$, $p \nmid m$, with eigenvalue $c(p)$. Then the representation π_f generated by the adelic lift Φ_f is irreducible. Let*

$$\pi_f = \otimes \pi_{f,p}$$

be the decomposition of this automorphic representation into local components.

Then:

   i) At the archimedean place we have

$$\pi_{f,\infty} = \pi^+_{m,k},$$

   the positive discrete series representation of weight k.

  ii) At a finite place $p \nmid 2m$ the local component $\pi_{f,p}$ is a spherical principal series representation

$$\pi_{f,p} = \pi_{\chi,m},$$

   where the unramified character $\chi : \mathbb{Q}_p^* \to \mathbb{C}$ is characterized by

$$c(p) = p^{k-3/2}(\chi(p) + \chi(p)^{-1}).$$

Proof: We proceed as was explained above, and decompose π_f into irreducibles:

$$\pi_f = \bigoplus_i \pi_i. \tag{7.27}$$

Each irreducible component is further decomposed into local representations:

$$\pi_i = \bigotimes_{p \le \infty} \pi_{i,p}.$$

Then we will invoke our multiplicity-one theorem 7.5.4 to show that all π_i are identical, which means there is only one i.

First we consider the archimedean components $\pi_{i,\infty}$. By Proposition 7.4.5, the differential operators $\mathcal{L}_Z$, $\mathcal{L}_{X_-}$, $\mathcal{L}_{X_+}$ have the following eigenvalues when applied to the adelic lift $\Phi = \Phi_f$:

$$\mathcal{L}_Z\Phi = k\Phi \qquad \mathcal{L}_{X_-}\Phi = \mathcal{L}_{Y_-}\Phi = 0.$$

Each $\pi_{i,\infty}$ then contains a vector with exactly the same properties. By our discussion in Section 3.1, this characterizes the discrete series representation $\pi^+_{m,k}$. Hence all $\pi_{i,\infty}$ are isomorphic, as we wanted to show. Moreover, i) is already proved.

Now we treat the finite places. Put $\Phi_f = \sum \Phi_i$ according to the decomposition (7.27). From $\pi_f(k)\Phi_f = \Phi_f$ for $k \in G^J(\mathbb{Z}_p)$ (see Proposition 7.4.5), it follows that for all i

$$\pi_i(k)\Phi_i = \Phi_i \qquad \text{for all } k \in G^J(\mathbb{Z}_p).$$

So it is clear that every local component π_p of π at a finite place p contains a non-zero $G^J(\mathbb{Z}_p)$-invariant vector, which means it is spherical. One of the hypotheses in ii) of Theorem 7.5.4 is therefore fulfilled.

Assume a *good place* $p \nmid 2m\infty$. By Theorem 6.3.10, the local representation $\pi_{i,p}$ is induced from an unramified character χ_i (resp., a subquotient of an induced representation). We have computed the Hecke eigenvalue of a K^J-invariant vector of such a representation in Theorem 6.4.6; it is given by

$$p^{3/2}\Big(\chi_i(p) + \chi_i(p)^{-1}\Big).$$

This is the eigenvalue under the operator $T^J(p)$, and from the connection between $T^J(p)$ and $T_{EZ}(p)$ in Proposition 6.6.5 we deduce the relation

$$c(p) = p^{k-3/2}\Big(\chi_i(p) + \chi_i(p)^{-1}\Big).$$

This proves that the $\pi_{i,p}$ are isomorphic for all i, and at the same time proves ii).

We see that condition i) of Theorem 7.5.4 is fulfilled if we let S be the set of primes dividing $2m$. It remains to check condition ii). Let p be a prime dividing m. Since f is an eigenform for W_p, we have

$$W_p f = \varepsilon f \qquad \text{for some } \varepsilon \in \{-1, 1\}.$$

By Proposition 7.4.8, we also have

$$\Phi * B_p = \varepsilon \Phi.$$

It follows that

$$\Phi_i * B_p = \varepsilon \Phi_i \qquad \text{for all } i.$$

This just means that the Heisenberg involution acts on the space of $G^J(\mathbb{Z}_p)$-invariant vectors in the local representation $\pi_{i,p}$ by the sign ε, for all i. This is what is required for ii) of Theorem 7.5.4 to be fulfilled. There is only one place which might not have been treated yet: The place 2 if $2 \nmid m$. But in this case the Heisenberg involution at 2 is by definition just the identity element of the local Hecke algebra. Consequently it acts by 1 on any irreducible representation, so that the condition in ii) of Theorem 7.5.4 is fulfilled in any case. This completes the proof. $\square$

In the course of the proof we have also seen:

7.5.6 Corollary. *Two Jacobi forms $f_1, f_2 \in J_{k,m}^{\mathrm{cusp}}$ generate the same automorphic G^J-representation if and only if they share the same eigenvalues for all $T^J(p)$, $p \nmid m$, and W_p, $p \mid m$.*

There is another corollary on Jacobi forms which is not easy to get by purely classical methods. It was first stated by Skoruppa in his thesis [Sk1].

7.5.7 Corollary. *There are no non-zero cuspidal Jacobi forms of weight one.*

Proof: Assume the converse, that is, some space $J_{1,m}^{\mathrm{cusp}}$ is non-zero. Since Hecke operators $T^J(p)$ and W_p for different primes commute, we can find a non-zero eigenform $f \in J_{1,m}^{\mathrm{cusp}}$. Let π_f the automorphic representation associated to f by Theorem 7.5.5. Let $\pi_{f,\infty}$ be the local component of π_f at infinity. By i) of the theorem, we have

$$\pi_{f,\infty} = \pi_{m,1}^+ = \pi_{sw}^m \otimes \tilde{\pi}_{1/2}^+.$$

The metaplectic automorphic representation $\tilde{\pi}$ corresponding to π_f by Corollary 7.3.5 therefore has local component $\tilde{\pi}_{1/2}^+$ at infinity. But by Corollary 3.2.7, this is a positive (even) Weil representation π_w^{m+}. We have thus arrived at a contradiction to Proposition 23 on p. 80 of [Wa1], which asserts that a cuspidal automorphic metaplectic representation cannot have a positive Weil representation as a local component at any place.   □

7.5.8 Remark. a) There is a version of Theorem 7.5.5 for skew-holomorphic Jacobi forms $f \in J_{k,m}^{*,\mathrm{cusp}}$ (see Section 4.1). The proof remains the same. We have only to replace in i) the positive discrete series representation $\pi_{m,k}^+$ by its counterpart $\pi_{m,k}^-$. The corollaries also remain valid for skew-holomorphic forms.

b) The "main theorem on Jacobi forms" by Skoruppa and Zagier ([SZ] Theorem 5 and its generalization mentioned in [Sk2], Section 6) yields a Hecke equivariant embedding of the space of Jacobi forms of weight k and index m into the space of ordinary elliptic modular forms of weight $2k-2$ and level m. One can use this theorem together with the Ramanujan-Petersson conjecture on the growth of Fourier coefficients of elliptic modular forms to improve statement ii) of Theorem 7.5.5: The spherical principal series representations $\pi_{\chi,m}$ mentioned there must necessarily belong to the continuous series, i.e., we have $|\chi(p)| = 1$.

c) It is a natural question to ask what are the local components of the representation π_f in Theorem 7.5.5 at the places $p|2m$. In fact, this is a serious problem also in the case of elliptic modular forms, where information is first of all given only at primes not dividing the level (see [Ge1] §5 C for more precise information). The above mentioned "main theorem on Jacobi forms" indicates that the two problems for Jacobi forms and elliptic modular forms are not independent. This turns out to be true, and becomes apparent once a representation theoretic version of the Skoruppa-Zagier map is established, as will be done in [Sch2]. A careful analysis of spherical representations of G^J at the bad places then not only yields information about Jacobi forms, but also about ordinary elliptic modular forms.

Bibliography

[Ar]   ARAKAWA, T.: *Real Analytic Eisenstein Series for the Jacobi Group.* Abh. Math. Sem. Univ. Hamburg **60** (1990), 131–148

[BB]   BAILY, W.L., JR., BOREL, A.: *Compactification of Arithmetic Quotients of Bounded Symmetric Domains.* Ann. of Math. 84 (1966), 442–528

[Be1]   BERNDT, R.: *Some Differential Operators in the Theory of Jacobi Forms.* IHES/M/84/10

[Be2]   BERNDT, R.: *Die Jacobigruppe und die Wärmeleitungsgleichung.* Math. Z. **191** (1986), 351–361

[Be3]   BERNDT, R.: *The Continuous Part of $L^2(\Gamma^J \backslash G^J)$ for the Jacobi Group G^J.* Abh. Math. Sem. Univ. Hamburg **60** (1990), 225–248

[Be4]   BERNDT, R.: *On the Characterization of Skew-holomorphic Jacobi Forms.* Abh. Math. Sem. Univ. Hamburg **65** (1995)

[Be5]   BERNDT, R.: *On Automorphic Forms for the Jacobi Group.* Jb. d. Dt. Math.-Verein. **97** (1995) 1–18

[Be6]   BERNDT, R.: *On Local Whittaker models for the Jacobi Group of Degree One.* Manuscripta Math. **84** (1994), 177–191

[BeBö]   BERNDT, R., BÖCHERER, S.: *Jacobi Forms and Discrete Series Representations of the Jacobi Group.* Math. Z. **204** (1990), 13–44

[B]   BOREL, A.: *Introduction to Automorphic Forms.* Proc. Symp. Pure Math., Vol. **9** (1966), 199–210

[BC]   BOREL, A., CASSELMAN, W. (ed.): *Automorphic Forms, Representations and L-functions,* Proc. Symp. Pure Math., Vol. **33**, Providence, Rhode Island 1979

[Bo]   BORHO, W.: *Primitive und vollprimitive Ideale in Einhüllenden von* $\mathfrak{sl}(5,\mathbb{C})$ J. Algebra **43** (1976), 619–654

[Bu]   BUMP, D.: *Automorphic Forms and Representations.* Cambridge University Press 1997

[Ca]   CARTIER, P.: *Representations of $\mathfrak{p}$-adic Groups: A Survey.* Proc. Symp. Pure Math., Vol. **33** (1979), part 1, 111–155

[Du]   DULINSKI, J.: *L-Functions for Jacobi forms on* $\mathbf{H} \times \mathbb{C}$. Result. Math. **31** (1997), 75–94

[EZ]   EICHLER, M., ZAGIER, D.: *The Theory of Jacobi Forms.* Birkhäuser, Boston 1985

[F]   FLATH, D.: *Decomposition of Representations into Tensor Products.* Proc. Symp. Pure Math., Vol. **33** (1979), part 1, 179–183

[Ge1]   GELBART, S.: *Automorphic Forms on Adele Groups.* Ann. math. studies **83**, Princeton 1975

[Ge2]   GELBART, S.: *Weil's Representation and the Spectrum of the Metaplectic Group.* Springer Lecture Note **530**, Berlin 1976

[GePS1] GELBART, S., PIATETSKI-SHAPIRO, I.: *On Shimura's Correpondence for Modular Forms of Half-integral Weight.* in: Proceedings, Colloquium on Automorphic Forms, Representation Theory and Arithmetic, Bombay 1979, Tata Institute of Fundamental Research Studies in Mathematics 10, Springer 1981

[GePS2] GELBART, S., PIATETSKI-SHAPIRO, I.: *Distinguished Representations and Modular Forms of Half-integral Weight.* Inv. Math. **59** (1980), 145–188

[GePS3] GELBART, S., PIATETSKI-SHAPIRO, I.: *Some Remarks on Metaplectic Cusp Forms and the Correspondences of Shimura and Waldspurger.* Israel J. of Math. **44** (1983), 97–126

[GGP]   GELFAND, I.M., GRAEV, M.I., PYATETSKII-SHAPIRO, I.I.: *Representation Theory and Automorphic Functions.* Academic Press, Boston 1990

[Go1]   GODEMENT, R.: *The Spectral Decomposition of Cusp Forms.* Proc. Symp. Pure Math. **9** (1966), 225–234

[Go2]   GODEMENT, R.: *Notes on Jacquet-Langlands' Theory.* Princeton 1970

[Gr]   GRITSENKO, V.: *Modulformen zur Paramodulgruppe und Modulräume der abelschen Varietäten.* Mathematica Gottingensis, Heft **12** (1995)

[HC]   HARISH-CHANDRA: *Automorphic Forms on Semisimple Lie Groups.* Springer Lecture Notes **62** (1968)

[He1]   HELGASON, S.: *Differential Operators on Homogeneous Spaces.* Acta Math. **102** (1959), 239–299

[He2]   HELGASON, S.: *Invariant Differential Operators and Eigenspace Representations.* in: Representation Theory of Lie Groups, Oxford 1977, Lond. Math. Soc. Lect. Note Series **34** (1979)

[Ho]   HOMRIGHAUSEN, J.: *Zur Darstellungstheorie der Jacobigruppe.* Dissertation. Hamburg 1995

[Ja]   JACQUET, H.: *Fonctions de Whittaker associées aux groupes de Chevalley.* Bull. Soc. Math. France **95** (1967), 243–309

[JL]   JACQUET, H., LANGLANDS, R.P.: *Automorphic Forms on* GL(2). Springer Lecture Notes **114** (1970)

[Ki]   KIRILLOV, A.A.: *Elements of the Theory of Representations.* Springer Berlin 1976

[Kn]   KNAPP, A.W.: *Representation Theory of Semisimple Groups.* Princeton University Press, Princeton 1986

[Ko]   KOHNEN, W.: *Jacobi Forms and Siegel Modular Forms: Recent Results and Problems.* L'Enseignement Math. **39** (1993), 121–136

[Kr]   KRAMER, J.: *An Arithmetic Theory of Jacobi Forms in Higher Dimension.* J. reine angew. Math. **458** (1995), 157–182

[Ku1]   KUBOTA, T.: *Topological Covering of* SL(2) *over a Local Field.* J. Math. Soc. Japan **19**, 1967, 114–121

[Ku2]   KUBOTA, T.: *Elementary Theory of Eisenstein Series.* John Wiley and Sons, New York 1973

[La]   LANG, S.: $SL_2(\mathbb{R})$. Springer, New York, 1985

[LV]   LION, G, VERGNE, M: *The Weil Representation, Maslov Index and Theta Functions.* Birkhäuser, Boston 1980

[Ma1]   MACKEY, G.: *Unitary Representations of Group Extensions 1.* Acta Math. **99** (1958), 265–311

[Ma2]   MACKEY, G.: *Unitary Group Representations in Physics, Probability and Number Theory.* Mathematical Lecture Notes Series **55**, Benjamin/Cummings Publ. 1978

[MM]   MATSUSHIMA, Y., MURAKAMI, S.: *On Vector Bundle Valued Harmonic Forms and Automorphic Forms on Symmetric Riemannian Manifolds.* Ann. Math. **78** (1963), 365–416

[Mum]   MUMFORD, D.: *Tata Lectures on Theta III.* Birkhäuser, Boston 1991

[Mu]   MURASE, A.: *L-Functions Attached to Jacobi Forms of Degree n. Part I: The Basic Identity.* J. reine ang. Math. **401** (1989), 122–156

[MW]   MOEGLIN, C., WALDSPURGER, J.-L.: *Décomposition Spectrale et Séries d'Eisenstein.* Birkhäuser 1994

[MVW]   MOEGLIN, C., VIGNÉRAS, M.-F., WALDSPURGER, J.-L.: *Correspondance de Howe sur un corps p-adique.* Springer Lecture Notes **1291** (1987)

[PS1]   PYATETSKI-SHAPIRO, I.: *Automorphic Functions and the Geometry of Classical Domains.* Gordon Breach, New York 1969

[PS2]   PYATETSKI-SHAPIRO, I.: *On the Saito-Kurokawa-Lifting.* Inv. Math. **71** (1983), 309–338

[Rao]   RAO, R. RANGA: *On some Explicit Formulas in the Theory of Weil Representation.* Pac. J. of Math. **157** (1993), 335–371

[Sa1]   SATAKE, I.: *Factors of Automorphy and Fock Representations.* Adv. in Math. **7** (1971), 83–110

[Sa2]   SATAKE, I.: *Algebraic Structures of Symmetric Domains.* Princeton University Press 1980

[Sch1] SCHMIDT, R.: *Über Schrödinger-Weil-Darstellungen und Thetafunktionen.* Hamburger Beiträge zur Mathematik, Heft **35**, 1995

[Sch2] SCHMIDT, R.: *Über automorphe Darstellungen der Jacobigruppe.* Dissertation, Hamburg 1998 (to appear).

[Sk1]   SKORUPPA, N.-P.: *Über den Zusammenhang zwischen Jacobiformen und Modulformen halbganzen Gewichts.* Dissertation, Bonner Math. Schriften Nr. **159**, Bonn 1985

[Sk2]   SKORUPPA, N.-P.: *Developments in the Theory of Jacobi Forms.* Bonn, MPI 89-40, or: Conference on Automorphic Forms and their Applications. Khabarovsk 1988, ed. by Kuznetsov (1990), 165–185

[Su]   SUGANO, T.: *Jacobi Forms and the Theta Lifting.* Comm. Math. Univ. St. Pauli **44** (1995), 1–58

[SZ]   SKORUPPA, N.-P., ZAGIER, D.: *Jacobi Forms and a Certain Space of Modular Forms.* Inv. Math. **94** (1988), 113–146

[Ta]   TAKASE, K.: *A Note on Automorphic Forms.* J. reine ang. Math. **409** (1990), 138–171

[Tate] TATE, J.: *Fourier Analysis in Number Fields and Hecke's Zeta Functions.* In: Cassels, Fröhlich: Algebraic Number Theory, Academic Press 1967

[Wa1] WALDSPURGER, J.-L.: *Correspondance de Shimura.* J. Math. pures et appl. **59** (1980), 1–133

[Wa2] WALDSPURGER, J.-L.: *Formes modulaires de poids demi-entier.* J. Math. pures et appl. **60** (1981), 375–484

[Wa3] WALDSPURGER, J.-L.: *Correspondance de Shimura et quaternions.* Forum Math. **3** (1991), 219–307

[We] WEIL, A.: *Sur certaines groupes d'opérateurs unitaires.* Acta Math. **111** (1964), 143–211

[WW] WHITTAKER, E.T., WATSON, G.N.: *A Course of Modern Analysis.* Cambridge University Press 1978

Index of Notations

Subgroups of G^J

Lie algebra notations and differential operators

p-adic representations and their spaces

Hecke theory

Miscellaneous notations

Index